Digestive Enzymes in Health and Disease

Digestive Enzymes in Health and Disease

Guest Editors

Stefan G. Pierzynowski
Kateryna Pierzynowska
Piotr Wychowański
Janine Donaldson

MDPI

Basel • Beijing • Wuhan • Barcelona • Belgrade • Novi Sad • Cluj • Manchester

Guest Editors

Stefan G. Pierzynowski
Department of Biology
Lund University
Lund
Sweden

Kateryna Pierzynowska
Department of Biology
Lund University
Lund
Sweden

Piotr Wychowański
Department of Head and
Neck and Sensory Organs
Catholic University of the
"Sacred Heart"
Rome
Italy

Janine Donaldson
Department of Physiology
University of the
Witwatersrand
Johannesburg
South Africa

Editorial Office
MDPI AG
Grosspeteranlage 5
4052 Basel, Switzerland

This is a reprint of the Special Issue, published open access by the journal *Biomolecules* (ISSN 2218-273X), freely accessible at: https://www.mdpi.com/journal/biomolecules/special_issues/B55D1TJFMV.

For citation purposes, cite each article independently as indicated on the article page online and as indicated below:

Lastname, A.A.; Lastname, B.B. Article Title. *Journal Name* **Year**, *Volume Number*, Page Range.

ISBN 978-3-7258-7354-8 (Hbk)
ISBN 978-3-7258-7355-5 (PDF)
https://doi.org/10.3390/books978-3-7258-7355-5

Contents

Preface

The Biomolecules Special Issue "Digestive Enzymes in Health and Disease" brings together various scientific contributions that highlight the complex roles of digestive enzymes in human physiology and pathology. Digestive enzymes are crucial for overall metabolic balance, ensuring efficient nutrient digestion and absorption, and maintaining the integrity of the gastrointestinal environment. Dysregulation of digestive enzyme function contributes to the development of a variety of clinical conditions. This Reprint provides readers with an updated exploration of these enzymes—extending beyond their classical digestive functions to include emerging "extra-digestive" roles, including new regulatory pathways, novel enzymatic functions, and mechanisms through which these enzymes contribute to disease progression or protection, thus deepening overall scientific understanding of digestive enzyme function and stimulating further inquiry into diagnostic and therapeutic innovations for metabolic complications resulting from enzyme dysfunction.

This Reprint is intended for a broad scientific and clinical audience, including (but not limited to) the following: Researchers in the fields of biochemistry, molecular biology, gastroenterology, and metabolism; clinicians treating gastrointestinal, metabolic, or inflammatory diseases; and translational scientists developing enzyme-targeted therapies. This Special Issue Reprint thus offers a platform for multidisciplinary dialogue and highlights areas in which enzyme-based interventions may soon transform patient care.

It is our hope that the works within this Reprint will inspire continued exploration into the biological significance of digestive enzymes and promote further investigations, thus fostering a deeper appreciation of the complexity, adaptability, and therapeutic promise of these essential biomolecules.

Stefan G. Pierzynowski, Kateryna Pierzynowska, Piotr Wychowański, and Janine Donaldson
Guest Editors

biomolecules

MDPI

Review

Symphony of Digestion: Coordinated Host–Microbiome Enzymatic Interplay in Gut Ecosystem

Volodymyr I. Lushchak [1,2]

[1] Department of Biochemistry and Biotechnology, Vasyl Stefanyk Precarpathian National University, Ivano-Frankivsk, 57 Shevchenko Str., 76018 Ivano-Frankivsk, Ukraine; volodymyr.lushchak@pnu.edu.ua

[2] Research and Development University, Ivano-Frankivsk, 13a Shota Rustaveli Str., 76018 Ivano-Frankivsk, Ukraine

Abstract: Digestion was once viewed as a host-driven process, dependent on salivary, gastric, pancreatic, and intestinal enzymes to break down macronutrients. However, new insights into the gut microbiota have redefined this view, highlighting digestion as a cooperative effort between host and microbial enzymes. Host enzymes initiate nutrient breakdown, while microbial enzymes, especially in the colon, extend this process by fermenting resistant polysaccharides, modifying bile acids, and transforming phytochemicals and xenobiotics into bioactive compounds. These microbial actions produce metabolites like short-chain fatty acids, which influence gut barrier function, immune regulation, and metabolism. I propose two frameworks to describe this interaction: the "duet," emphasizing sequential enzymatic cooperation, and the "orchestra," reflecting a spatially and temporally coordinated system with host–microbiota feedback. Disruption of this symbiosis, through antibiotics, inflammation, diet, or aging, leads to dysbiosis, impaired digestion, and contributes to metabolic, neurologic, cardiovascular, and inflammatory diseases. Recognizing digestion as a dynamic, integrated system opens new paths for therapies and nutrition. These include enzyme-targeted prebiotics, probiotics, postbiotics, and personalized diets. Embracing this systems-level perspective enables innovative diagnostics and treatments, aiming to restore enzymatic balance and improve digestive and systemic health.

Keywords: gut microbiota; digestive enzymes; nutrient sensing; host–microbiome interaction; intestinal barrier; oxidative stress; microbial metabolism; precision nutrition; hormonal regulation; short-chain fatty acids (SCFAs)

1. Introduction

Digestion is a fundamental biological process that enables the breakdown of complex food matrices into absorbable nutrients essential for human survival and health. Traditionally, the gastrointestinal tract (GIT) has been viewed primarily as a mechanical and enzymatic system orchestrated by the host's secretions, including salivary amylase, gastric pepsin, pancreatic amylase, proteases, lipases, and a suite of brush-border enzymes. This classical perspective posits that human digestive enzymes dominate in nutrient processing, functioning largely independently of other biological factors. However, advances in microbial ecology, molecular biology, and high-throughput sequencing have profoundly reshaped our understanding of this intricate system. The gut microbiota—the vast and diverse community of microorganisms residing in the digestive tract—has emerged as a pivotal player in digestion, contributing a vast arsenal of enzymes capable of degrading dietary components that the host enzymes cannot efficiently process on their own [1–3].

The emerging paradigm recognizes the gut as a complex and dynamic ecosystem, where digestive processes result from a coordinated interplay between host-derived enzymes and microbial enzymatic activities. Rather than working in isolation or redundancy, these enzymes form a synergistic network that dynamically adapts specific genotypes to diet, health status, and environmental factors. Such complicated cooperation may be described as a musical performance, where the host and microbiota each contribute their own "instruments" to a harmonious "orchestra," or at times engage in a more intimate "duet," complementing and completing each other's functions [4–6]. For example, human enzymes initiate the breakdown of starches and proteins, while microbial enzymes further ferment resistant carbohydrates and complex polyphenols, producing short-chain fatty acids (SCFAs) and other bioactive metabolites critical for colonic health and systemic metabolism [7,8].

Understanding this coordinated enzymatic symphony is crucial because disturbances in this relationship are linked to a range of pathologies, including inflammatory bowel disease, obesity, metabolic syndrome, and neurodegenerative disorders [9]. Moreover, the integration of host and microbial enzymatic contributions opens new avenues for therapeutic interventions, nutritional strategies, and personalized medicine aimed at restoring or optimizing digestive health. This review explores the enzymatic roles of the host and the gut microbiota in digestion, conceptual models of their cooperation, the consequences of their dysregulation, and prospects for targeting this synergy to promote health. Such a systems approach paves the way for precision nutrition, offering a powerful tool for personalized prevention and treatment of most non-communicable diseases.

2. The Enzymatic Roles of the Host and the Gut Microbiota

The human gastrointestinal tract operates as a highly coordinated system in which both host-derived and microbiota-derived enzymes play essential roles in the digestion and biotransformation of nutrients. Host enzymes are primarily responsible for the initial stages of macronutrient breakdown, while microbial enzymes complement and extend this process by degrading complex dietary components that escape host digestion, including dietary fibers, polyphenols, and resistant starches. This enzymatic interplay significantly influences nutrient absorption, energy harvest, immune responses, and gut barrier integrity. Table 1 summarizes the key enzymes produced by the host and gut microbiota, highlighting their substrates, products, and physiological significance.

Table 1. Comparison of host and microbial digestive enzymes by substrate and function.

Substrate	Host-Derived Enzymes	Microbial Enzymes	Primary Function
Starches and simple carbohydrates	Salivary amylase, pancreatic α-amylase	Amylases, pullulanases, maltases, isomaltases (CAZymes)	Hydrolysis of starch and oligosaccharides into monosaccharides
Non-digestible polysaccharides	None	Cellulases, hemicellulases, xylanases, pectinases	Fermentation of dietary fibers; SCFA production
Disaccharides	Sucrase, maltase, lactase (brush-border enzymes)	β-Galactosidase, β-fructofuranosidase	Breakdown of sucrose, maltose, and lactose into monosaccharides
Proteins and peptides	Pepsin, trypsin, chymotrypsin, peptidases	Microbial proteases, peptidases	Peptide hydrolysis; bioactive amine generation
Fats and lipids	Gastric and pancreatic lipases, phospholipases	Lipases, esterases	Hydrolysis of triglycerides and other lipids; production of microbial lipid metabolites

Table 1. *Cont.*

Substrate	Host-Derived Enzymes	Microbial Enzymes	Primary Function
Bile acids	Liver secretion and reabsorption control	Bile salt hydrolases, hydroxysteroid dehydrogenases	Deconjugation and transformation of bile acids; host–microbiome signaling
Polyphenols and phytochemicals	Limited phase I/II metabolism (e.g., glucuronidation)	Glycosidases, esterases, decarboxylases	Microbial conversion into bioavailable phenolic compounds
Xenobiotics and drugs	Cytochrome P450s, conjugating enzymes (UGT, SULT, etc.)	Reductases, azoreductases, β-glucuronidases	Biotransformation, reactivation, or detoxification of xenobiotics

Abbreviations: CAZymes: carbohydrate-active enzymes; SCFAs: short-chain fatty acids; UGT: uridine 5′-diphospho-glucuronosyltransferase; SULT: sulfotransferase.

2.1. Host Digestive Enzymes

The host digestive system secretes a complex array of enzymes that initiate and execute the hydrolysis of macronutrients. Digestion begins in the oral cavity with salivary amylase, which initiates starch breakdown into maltose and dextrins. The acidic environment of the stomach activates pepsinogen to pepsin, which denatures proteins and cleaves peptide bonds, facilitating further proteolysis in the small intestine [3–5]. The pancreas secretes a cocktail of digestive enzymes, including amylases, lipases, proteases (trypsin, chymotrypsin), and nucleases, which collectively break down carbohydrates, lipids, proteins, and nucleic acids. Finally, brush-border enzymes such as maltase, lactase, and sucrase complete the digestion of disaccharides into monosaccharides suitable for absorption [5,10].

The secretion and activity of these enzymes are tightly regulated by neural signals; hormonal controls such as gastrin, cholecystokinin, and secretin; and influenced by nutrient sensing mechanisms that modulate digestive capacity in response to diet composition and feeding status [11,12]. This adaptive regulation ensures efficient digestion under varying physiological conditions and dietary intakes. However, despite this sophisticated host enzymatic machinery, many complex polysaccharides and other dietary components remain indigested without microbial assistance.

2.2. Microbial Digestive Enzymes

The human gut microbiota encodes an extensive repertoire of carbohydrate-active enzymes (CAZymes) not produced by the host, including glycoside hydrolases, polysaccharide lyases, and carbohydrate esterases that degrade a wide range of complex polysaccharides such as resistant starches, cellulose, pectins, and hemicelluloses [1,7,8,13,14]. Additionally, microbial proteases complement host proteolysis by breaking down dietary and endogenous proteins in the colon, releasing peptides and amino acids that may serve as substrates for further microbial metabolism [15]. Bile salt hydrolases, another class of microbial enzymes, modulate bile acid profiles by deconjugating bile salts, thus influencing lipid digestion, cholesterol metabolism, and intestinal signaling pathways [16,17]. The microbial fermentation of undigested carbohydrates generates SCFAs—primarily acetate, propionate, and butyrate—that serve as energy sources for colonocytes, regulate immune responses, and impact systemic metabolism through signaling mechanisms [1,4,8,13,14,18]. Interestingly, microbes not only produce bioactive metabolites such as vitamins (K, B12), neurotransmitter precursors, and phenolic derivatives, highlighting their broad enzymatic influence beyond mere digestion, but also use them [19–21]. In addition to broadly discussed items, there are some specific routes for the transformation of specific compounds by the gut microbiota and which may substantially affect hosts. For example, gut microbiota may convert glucoraphanin, a glucosinolate found in cruciferous plants, to isothiocyanate sulforaphane and related compounds [22].

Spatially, enzymatic activities vary significantly along the gut axis. The small intestine harbors fewer microbes, predominantly facultative anaerobes, whereas the colon hosts a dense and diverse anaerobic community with extensive polysaccharide-degrading capacity. This spatial organization reflects adaptation to substrate availability and environmental conditions, further emphasizing the complementary nature of host and microbial enzymes [4,10,12].

3. Models of Cooperation: Duet or Orchestra?

Conceptualizing the host–microbiome enzymatic interaction is essential to understanding digestion beyond a sum of parts. Two models frame this cooperation: the "duet" and the "orchestra" (Figure 1). In the duet model (Figure 1, left), enzymatic activities are sequential or complementary—host enzymes initiate digestion, and microbial enzymes complete it, and vice versa. For instance, salivary and pancreatic amylases hydrolyze starch to maltose and oligosaccharides, which escape absorption in the small intestine and are fermented by microbial amylases and glycoside hydrolases into SCFAs in the colon [1,7,13,18,23]. Similarly, the host may degrade dietary polyphenols partially, enabling microbial enzymes to generate bioactive metabolites with systemic effects [24,25].

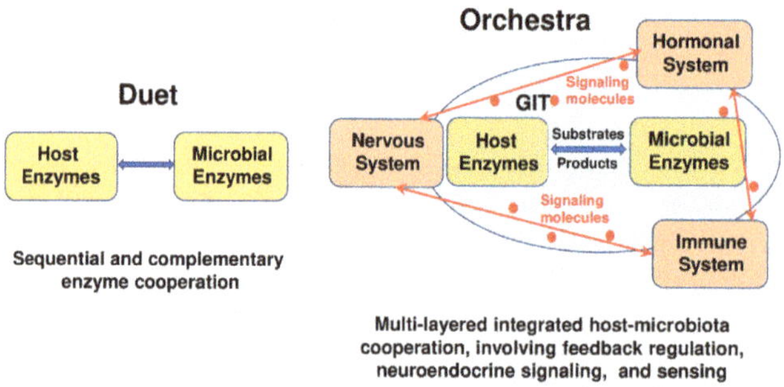

Figure 1. Conceptual model of enzymatic cooperation: "Duet" vs. "Orchestra". Description in text. Abbreviations: GIT: gastrointestinal tract.

The orchestra model (Figure 1, right) expands this concept to a highly coordinated, multi-layered interaction where host and microbial enzymes operate in a spatially and temporally regulated manner with feedback loops. For example, bile acids secreted by the liver are modified by microbial bile salt hydrolases, which modulate bile acid signaling via nuclear receptors (FXR, TGR5), influencing digestion, host metabolism, and immune responses [17,26,27]. This multidirectional crosstalk highlights an intricate network where enzymes, microbial populations, and host cells communicate dynamically.

Metagenomics and metabolomics studies reveal that gene expression profiles of microbial enzymes change in response to diet and host physiology, underscoring the system's plasticity [28,29]. Gnotobiotic animal models confirm that colonization by specific microbial consortia influences host enzyme expression and digestive efficiency [30,31].

4. Disruption of Enzymatic Harmony: Dysbiosis and Disease

The finely tuned enzymatic cooperation between host and microbiota can be disrupted by antibiotics, chronic inflammation, aging, and dietary changes, leading to dysbiosis—an imbalance in microbial composition and function [32–34]. Antibiotics reduce microbial diversity and enzyme pools, impairing carbohydrate fermentation and SCFA production, which negatively affects gut barrier integrity and immune homeostasis [32,35]. Inflammatory bowel diseases exhibit altered expression of microbial proteases, contributing to mucosal damage and nutrient malabsorption [36].

Aging is associated with reduced microbial diversity and enzymatic activity, contributing to diminished digestive efficiency, increased intestinal permeability, and systemic inflammation ("inflammaging") [21,32–34]. Diets low in fiber reduce substrates for microbial fermentation, leading to decreased SCFA levels and altered bile acid metabolism, which are linked to metabolic disorders such as obesity and type 2 diabetes [33,37].

The loss or overexpression of specific microbial enzymes—such as excessive proteolytic activity—can produce harmful metabolites, exacerbate inflammation, and impair nutrient absorption [34,38,39]. It clearly shows the importance of enzymatic balance for maintaining gut and systemic health.

5. Therapeutic Perspectives and Future Directions

Targeting the enzymatic synergy in the gut opens new therapeutic opportunities. Probiotics engineered or selected for specific enzyme activities ("enzyme-targeted probiotics") can restore deficient functions, enhance fiber degradation, or modulate bile acid metabolism [40,41]. Postbiotics—microbial metabolites or enzyme preparations—offer alternative means to supplement or regulate enzymatic activity [42,43].

Dietary modulation remains a cornerstone; prebiotic fibers selectively stimulate beneficial enzyme-expressing microbes, improving SCFA production and digestive efficiency. Personalized nutrition approaches integrating gut microbiota enzymatic profiles and host genetics promise tailored dietary recommendations for optimal digestive health.

Future research will leverage systems biology to model the complex enzymatic networks and synthetic biology to engineer microbial consortia with desired enzymatic functions. Microbiome engineering may enable the creation of designer ecosystems to prevent or treat digestive and metabolic diseases.

6. Some Poorly Covered or Overlooked Points

In the study of digestion, attention is mostly focused on carbohydrates, lipids, and proteins, while nucleic acids are much less covered. In the human GIT, nucleic acid digestion is a complex process involving both host and microbial enzymes that ensures the breakdown and absorption of DNA and RNA derived from diet, sloughed epithelial cells, and microbiota. Upon ingestion, nucleic acids are first denatured in the stomach by the acidic environment (pH ~1.5–3.5). It disrupts their secondary and tertiary structures, making them more accessible to enzymatic action. In the small intestine, pancreatic secretions containing deoxyribonuclease I (DNase I) and ribonuclease A (RNase A) hydrolyze DNA and RNA into smaller oligonucleotides. These are subsequently degraded by brush-border enzymes such as nucleotidases and nucleosidases into nucleosides and nitrogenous bases (e.g., adenine, guanine, cytosine, uracil, thymine), as well as pentose carbohydrates (ribose, deoxyribose). Specific nucleoside and base transporters in the enterocyte membrane, including concentrative nucleoside transporters (CNTs) and equilibrative nucleoside transporters (ENTs), facilitate their absorption into the portal circulation. However, not all nucleic acid components are absorbed in the small intestine. Residual nucleotides and nucleic acids pass into the colon, where they are metabolized by the gut microbiota. Certain microbial taxa, including *Bacteroides* spp., *Clostridium* spp., and *Faecalibacterium prausnitzii*, produce extracellular DNases and RNases, contributing to the degradation of residual nucleic acids [44,45].

The human gut microbiome harbors a diverse set of microbial species capable of expressing extracellular nucleases that degrade residual nucleic acids. These microbes utilize liberated nucleotides and bases in their purine and pyrimidine salvage pathways for DNA/RNA synthesis, or ferment nucleic acid-derived sugars into short-chain fatty acids (SCFAs) such as acetate, propionate, and butyrate—key molecules in host energy metabolism and intestinal health [41,46]. Microbial metabolism of nucleic acids also contributes to nitrogen recycling in the colon. Beyond energy metabolism, microbial nucleic acid turnover also affects host immune signaling; unmethylated CpG motifs from bacterial DNA can activate Toll-like receptor 9 (TLR9), initiating immune responses that may contribute to chronic intestinal inflammation [47,48]. Dysregulated microbial nucleic acid metabolism has been implicated in inflammatory bowel disease (IBD), where excessive CpG DNA sensing may disrupt mucosal tolerance, and moreover, imbalances in purine metabolism—potentially driven by microbial activity—have been linked to colorectal cancer through genotoxic stress and altered nucleotide pools, as well as metabolic disorders [49]. Furthermore, dietary nucleotides have been shown to influence gut development, promote beneficial microbial populations such as *Bifidobacterium*, and improve immune function in neonates and immunocompromised individuals [50]. As such, both endogenous and dietary nucleic acids, along with microbial contributions, play a crucial role in maintaining intestinal homeostasis and systemic health [51,52]. Dietary nucleotides can also modulate the gut microbiota, enhance mucosal immunity, and influence early gut development. Altogether, these observations highlight nucleic acid digestion as an overlooked but important contributor to the host–microbiota metabolic axis, with implications for immune regulation, gut barrier function, and disease pathogenesis.

Several points can be highlighted based on our previous experiments with GIT inhabitants. For example, different strains of the same bacterial species may exhibit varying sensitivity to environmental stresses. In the case of *Escherichia coli*, we observed that differences in oxygen tolerance among strains were associated with their distinct capacities to manage reactive oxygen species (ROS), particularly through the activity of antioxidant enzymes [53]. Exposure of bakery yeasts *Saccharomyces cerevisiae* to hydrogen peroxide triggered a hormetic reshaping of their phenotype [54]. Moreover, the specific pattern of this reorganization was influenced by the type of carbohydrate used during yeast cultivation. Finally, *S. cerevisiae*, under certain conditions, may produce polypeptides—enzyme precursors—which, under stress conditions, may mature to active forms [55]. Despite the above three works being carried out in in vitro conditions, they clearly show that many more parameters, as currently practiced, should be taken into consideration when dealing with such a complex GIT ecosystem.

7. Orchestrating Host–Gut Microbiota Cooperation: A System Conducting

The digestion of food is not a primitive, mechanical, or chemical process. It is a very complicated system where reactions carried out by the host and gut microbiota digestive enzymes are tightly regulated on short and long-term time scales. Both hormonal and neural regulations in concert govern this system. Similarly to an orchestra, an integrated network of neuronal and endocrine signals is coordinated by a conductor to synchronize the gut ecosystem. The gastrointestinal tract (GIT), often described as a "second brain," is equipped with extensive neural circuits and hormone-secreting cells that ensure the temporal and spatial coordination of digestion, absorption, and microbial activity [56,57]. This orchestrated regulation is essential for maintaining digestive efficiency and metabolic homeostasis in the host. The enteric nervous system (ENS), an intricate network of over 100 million neurons embedded within the gut wall, is the heart of such orchestration [57].

So, digestion is regulated through a complex interplay of neural signals from both the enteric (ENS) and central nervous systems (CNS), with communication between gut regions helping coordinate digestive processes. Enteric neurons use diverse chemical messengers that act on numerous receptors, offering multiple potential targets for modifying digestive function. The ENS autonomously regulates key functions such as peristalsis, secretion of digestive juices, vascular tone, and communication with immune and epithelial cells. Importantly, ENS neurons are equipped with receptors that sense microbial metabolites, including SCFAs, bile acids, and tryptophan derivatives. These microbial signals modulate neuronal activity and, in turn, affect gut motility, secretion rates, and mucosal barrier function [58]. Figure 2 schematically demonstrates host–gut microbiota cooperation via SCFAs. This bidirectional communication enables a responsive and flexible digestive environment that adapts to both dietary and microbial fluctuations.

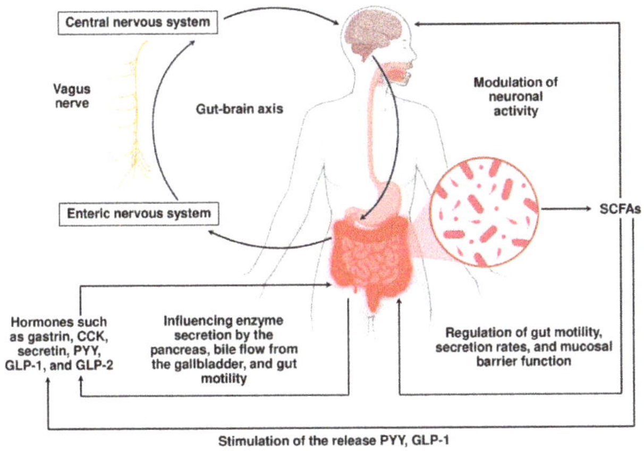

Figure 2. Orchestrating host–gut microbiota cooperation: plural feedback and forward signaling. The enteric and central nervous systems (ENS and CNS), connected via the vagus nerve, form a bidirectional communication system, the gut–brain axis, integrating signals from the CNS, ENS, and gut microbiota. SCFAs, such as acetate, propionate, and butyrate, are key microbial metabolites affecting gut physiology and influencing both the ENS and CNS via receptor-mediated signaling pathways, thereby modulating motility, barrier function, hormone release, and even behavior. In parallel, enteroendocrine hormones coordinate digestive functions, and enteric neurotransmitters convey gut-derived cues to the brain through vagal nerve signaling. This integrated network enables flexible adaptation to dietary and microbial fluctuations, while feedback loops maintain homeostasis in digestion, metabolism, and immune responses. Abbreviations: CCK—cholecystokinin; PYY—peptide YY; GLP-1/2—glucagon-like peptide 1/2; SCFAs—short-chain fatty acids. Figure partially created using BioRender.com (accessed on 25 July 2025). A range of gastrointestinal hormones, in parallel to the neural system, act as biochemical messengers coordinating the actions of digestive organs and relaying information to the CNS. Enteroendocrine cells, in response to luminal nutrients and microbial cues, secrete some hormones, and key among them are gastrin, cholecystokinin (CCK), secretin, peptide YY (PYY), and glucagon-like peptides (GLP-1 and GLP-2) [59,60]. These hormones influence enzyme secretion by the pancreas, bile flow from the gallbladder, and gut motility, while also modulating appetite and systemic glucose homeostasis. Of particular interest is the ability of microbial metabolites, especially SCFAs like butyrate and propionate, to stimulate the release of GLP-1 and PYY, thereby linking microbial fermentation to satiety and metabolic control [12]. The vagus nerve plays a critical role in connecting the gut and brain through the so-called gut–brain axis. Vagal afferent fibers transmit signals from the gut to the brain regarding nutrient content, hormone levels, and microbial metabolites, while efferent fibers regulate secretion, motility, and inflammatory responses [61]. The vagus nerve also mediates the effects of stress and emotional states on digestive physiology. Chronic stress, for example, can suppress vagal tone, reduce digestive secretions, and disrupt gut microbial composition, leading to a cascade of dysregulated processes. Conversely, interventions that enhance vagal activity, such as meditation, deep breathing, or electrical vagal stimulation, have been shown to improve gut function and microbial diversity.

The influence of the microbiota on neuroendocrine regulation extends beyond simple feedback. Microbes are capable of producing a variety of neuroactive substances, including gamma-aminobutyric acid (GABA), dopamine, serotonin precursors, and acetylcholine-like compounds. These substances can affect the ENS directly or signal to the CNS via the bloodstream and vagus nerve. Moreover, microbial metabolites interact with the hypothalamic–pituitary–adrenal (HPA) axis, influencing systemic stress responses, appetite regulation, and metabolic rate [61]. Those collectively show that the gut microbiota is not the only participant in digestion but also acts as an active endocrine organ with the capacity to influence host behavior and physiology on a systemic level.

An additional layer of complexity is introduced by the nutrient-sensing mechanisms embedded in both the host and microbial components of the gut. Enteroendocrine cells are equipped with receptors for glucose, amino acids, fatty acids, and microbial metabolites, allowing them to integrate diverse signals and modulate hormonal output accordingly. Similarly, microbes adjust their gene expression profiles—including the expression of carbohydrate-active enzymes (CAZymes)—in response to available substrates and host-secreted molecules [14,15]. This coordinated sensing system enables the host and microbiota to anticipate and respond to dietary inputs in a synchronized manner, optimizing digestion and nutrient utilization.

Disruption of this finely tuned orchestration can have profound consequences. Factors such as antibiotic treatment, poor diet, chronic stress, aging, and disease can impair neuroendocrine signaling, disturb microbial composition, and reduce enzymatic efficiency [32,35]. Dysregulated hormonal responses may lead to impaired satiety signaling, delayed gastric emptying, or excessive bile acid accumulation, each of which can alter microbial niches and exacerbate dysbiosis [30,62]. Likewise, vagal dysfunction may impair feedback mechanisms essential for maintaining motility and mucosal health. The resulting breakdown in coordination can contribute to a wide range of disorders, from irritable bowel syndrome and metabolic syndrome to mood disorders and neurodegeneration [9,20].

The therapeutic implications of these insights are substantial. Targeting neuroendocrine regulators and their interaction with the microbiota opens new possibilities for improving digestive and systemic health. Prebiotics and dietary fibers that enhance SCFA production can stimulate beneficial hormone release and modulate ENS activity [7,37]. Probiotic strains selected for their ability to produce neuroactive compounds may serve as "psychobiotics" to improve mood and cognition. Personalized nutrition strategies that consider individual hormonal profiles, stress levels, and microbiota composition could optimize digestive efficiency and metabolic outcomes [62,63]. Furthermore, emerging technologies such as bioelectronic medicine and synthetic biology offer the potential to modulate the gut–brain axis in real time.

In conclusion, digestion is not merely a chemical process but a symphony conducted by an intricate network of neural and hormonal signals, finely attuned to the metabolic and structural contributions of the gut microbiota. Understanding this orchestration enriches our view of digestion from a linear sequence of breakdown steps to a dynamically regulated, co-evolved system. By embracing this perspective, future research and clinical practice can move toward integrative approaches that preserve and restore harmony in the digestive orchestra.

Moreover, the circadian regulation of both host and microbial processes adds another axis of orchestration. Circadian rhythms, driven by central and peripheral clocks, influence feeding behavior, hormone secretion, gut motility, and microbial activity. Disruption of these rhythms, as seen in shift work, jet lag, or irregular eating patterns, leads to desynchronization of digestive processes and microbial dysbiosis. For example, it was demonstrated that the intestinal microbiota, in both mice and humans, exhibited diurnal oscillations that are influenced by feeding rhythms, leading to time-specific compositional and functional profiles over the course of a day [38]. Interestingly, disruption of the host circadian clock via genetic ablation or jet lag led to microbiota dysbiosis driven by impaired feeding rhythms, resulting in glucose intolerance and obesity that are transferable to germ-free mice, thus revealing a microbiome-dependent mechanism linking circadian disruption to metabolic disorders in humans [38]. Restoration of rhythmicity may be a promising strategy to optimize host–gut microbiota cooperation and improve metabolic outcomes in such cases.

A systems biology perspective, integrating data from transcriptomics, metabolomics, and microbiota profiling, is essential to fully understand the complex orchestration of digestion. Computational models simulating gut neuroendocrine–microbiota interactions can help predict the impact of dietary or pharmacologic interventions on enzyme activity, microbial composition, and host health. Such models can also inform the development of synthetic microbial communities or engineered probiotics tailored to specific host needs. Moving forward, personalized interventions targeting the neuroendocrine pathways, either through diet, microbiota modulation, or behavioral change, offer a powerful approach to restore symphony in the gut ecosystem and mitigate digestive, metabolic, and neuropsychiatric disorders.

An intriguing question remains: Who is the conductor of the orchestra? This question arose many times, but it was not answered. Individual fast responses are coordinated by the nervous system, and slow responses are coordinated by the hormonal system. Together, they form the neurohumoral system. But that is not the whole story. The neurohumoral system does not operate in isolation. It cooperates tightly with other systems, including both its components of the digestive system, the host and microbiome. How do they coordinate their operation? In a classic way, via the system of forward and (mainly) feedback signals. So, there is no single conductor, and the orchestra is conducted by several cooperating conductors. That together can be called a "collective wisdom".

8. Conclusions and Perspectives

In this review, we have highlighted the intricate relationships between gut microbiota, host digestive processes, and the regulation of the activities of enzymes under the influence of neural, hormonal, and nutritional cues. A growing body of evidence demonstrates that the gut microbiota not only metabolizes dietary components but also actively shapes host physiology through the production of signaling molecules, modulation of the intestinal barrier, and interactions with host immune and endocrine systems.

Key gastrointestinal hormones, including gastrin, cholecystokinin, and secretin, play pivotal roles in coordinating digestive enzyme secretion in response to both dietary and microbial cues. Furthermore, nutrient sensing mechanisms and the redox state of the gut environment critically affect microbial activity, enzyme expression, and barrier integrity. These dynamic interactions underscore the gut as a highly adaptive ecosystem that is affected by both intrinsic and extrinsic factors.

Despite significant progress, many aspects of gut microbiota–host communication remain poorly understood. Differences in microbial strain responses to oxidative stress, diet, and metabolic demands present both a challenge and an opportunity for deeper functional characterization. The emerging concept of hormesis in microbial responses to environmental stressors, including oxidative agents and toxicants, adds another layer of complexity that deserves further investigation.

Looking forward, future research should focus on the following areas:

- Strain-specific responses—elucidating how different microbial strains within the same species respond to oxidative and nutritional stresses will enhance our understanding of gut resilience and adaptability.
- Integration of multi-omics approaches—combining metagenomics, transcriptomics, metabolomics, and proteomics will provide comprehensive insights into the dynamic interplay between microbes, enzymes, and host responses.
- Personalized nutrition and gut microbiota modulation—advancing microbiota-targeted interventions, including prebiotics, probiotics, and postbiotics, tailored to individual microbiome compositions, could significantly improve metabolic and gastrointestinal health. Such an approach may be a cornerstone for precision nutrition, leveraging high-throughput sequencing, metabolomics, and bioinformatics to tailor dietary recommendations based on individual genetic profiles, microbiome composition, lifestyle, and environmental factors, aiming to optimize nutrient intake, promote a healthy microbiome, and reduce disease risk.
- Therapeutic targeting of the gut barrier—strengthening intestinal barrier function through dietary, microbial, or pharmacological means offers a promising avenue to counteract systemic inflammation and chronic diseases linked to gut dysbiosis.

Altogether, an integrated, systems-level understanding of the gut ecosystem—bridging microbiology, enzymology, physiology, and nutrition—will be crucial for developing next-generation strategies to support human health and mitigate gastrointestinal and metabolic disorders.

Funding: This research received no external funding.

Institutional Review Board Statement: Not applicable.

Informed Consent Statement: Not applicable.

Data Availability Statement: No new data were created.

Acknowledgments: The author appreciates T. Dmytriv. for carefully reviewing the manuscript and helping with the figures. AI was used to polish the text and identify some references.

Conflicts of Interest: The author declares no conflicts of interest.

Abbreviations

The following abbreviations are used in this manuscript:

B12	vitamin B12 (Cobalamin)
CAZymes	carbohydrate-active enzymes
GIT	gastrointestinal tract
H_2O_2	hydrogen peroxide
ROS	reactive oxygen species
SCFA	short-chain fatty acid
TLR	toll-like receptor

References

1. Rowland, I.; Gibson, G.; Heinken, A.; Scott, K.; Swann, J.; Thiele, I.; Tuohy, K. Gut Microbiota Functions: Metabolism of Nutrients and other Food Components. *Eur. J. Nutr.* **2018**, *57*, 1–24. [CrossRef]
2. Sun, Y.; Wang, X.; Li, L.; Zhong, C.; Zhang, Y.; Yang, X.; Li, M.; Yang, C. The Role of Gut Microbiota in Intestinal Disease: From an Oxidative Stress Perspective. *Front. Microbiol.* **2024**, *15*, 1328324. [CrossRef]
3. Jiang, Z.; Mei, L.; Li, Y.; Guo, Y.; Yang, B.; Huang, Z.; Li, Y. Enzymatic Regulation of the Gut Microbiota: Mechanisms and Implications for Host Health. *Biomolecules* **2024**, *14*, 1638. [CrossRef] [PubMed]
4. Karasov, W.H.; Diamond, J.M. Interplay between Physiology and Ecology in Digestion: Intestinal Nutrient Transporters Vary within and between Species According to Diet. *BioScience* **1988**, *38*, 602–611. [CrossRef]

5. Clark, J.A.; Coopersmith, C.M. Intestinal Crosstalk: A New Paradigm for Understanding the Gut as the "Motor" of Critical Illness. *Shock* **2007**, *28*, 384–393. [CrossRef] [PubMed]
6. Gregor, R.; Probst, M.; Eyal, S.; Aksenov, A.; Sasson, G.; Horovitz, I.; Dorrestein, P.C.; Meijler, M.M.; Mizrahi, I. Mammalian Gut Metabolomes Mirror Microbiome Composition and Host Phylogeny. *ISME J.* **2022**, *16*, 1262–1274. [CrossRef]
7. Deehan, E.C.; Duar, R.M.; Armet, A.M.; Perez-Muñoz, M.E.; Jin, M.; Walter, J. Modulation of the Gastrointestinal Microbiome with Nondigestible Fermentable Carbohydrates to Improve Human Health. *Microbiol. Spectr.* **2017**, *5*, 10.1128. [CrossRef]
8. de Vos, W.M.; Tilg, H.; Van Hul, M.; Cani, P.D. Gut Microbiome and Health: Mechanistic Insights. *Gut* **2022**, *71*, 1020–1032. [CrossRef]
9. Shabani, M.; Ghoshehy, A.; Mottaghi, A.M.; Chegini, Z.; Kerami, A.; Shariati, A.; Taati Moghadam, M. The Relationship between Gut Microbiome and Human Diseases: Mechanisms, Predisposing Factors and Potential Intervention. *Front. Cell. Infect. Microbiol.* **2025**, *15*, 1516010. [CrossRef]
10. Sensoy, I. A Review on the Food Digestion in the Digestive Tract and the Used in vitro Models. *Curr. Res. Food Sci.* **2021**, *4*, 308–319. [CrossRef]
11. Lu, V.B.; Gribble, F.M.; Reimann, F. Nutrient-Induced Cellular Mechanisms of Gut Hormone Secretion. *Nutrients* **2021**, *13*, 883. [CrossRef]
12. Cho, H.; Lim, J. The Emerging Role of Gut Hormones. *Mol. Cells* **2024**, *47*, 100126. [CrossRef]
13. Flint, H.J.; Scott, K.P.; Duncan, S.H.; Louis, P.; Forano, E. Microbial Degradation of Complex Carbohydrates in the Gut. *Gut Microbes* **2012**, *3*, 289–306. [CrossRef]
14. Wardman, J.F.; Bains, R.K.; Rahfeld, P.; Withers, S.G. Carbohydrate-Active Enzymes (CAZymes) in the Gut Microbiome. *Nat. Rev. Microbiol.* **2022**, *20*, 542–556. [CrossRef] [PubMed]
15. Diether, N.E.; Willing, B.P. Microbial Fermentation of Dietary Protein: An Important Factor in Diet–Microbe–Host Interaction. *Microorganisms* **2019**, *7*, 19. [CrossRef] [PubMed]
16. Chand, D.; Avinash, V.S.; Yadav, Y.; Pundle, A.V.; Suresh, C.G.; Ramasamy, S. Molecular Features of Bile Salt Hydrolases and Relevance in Human Health. *Biochim. Biophys. Acta Gen. Subj.* **2017**, *1861 Pt A*, 2981–2991. [CrossRef] [PubMed]
17. Dong, Z.; Yang, S.; Tang, C.; Li, D.; Kan, Y.; Yao, L. New Insights into Microbial Bile Salt Hydrolases: From Physiological Roles to Potential Applications. *Front. Microbiol.* **2025**, *16*, 1513541. [CrossRef]
18. den Besten, G.; van Eunen, K.; Groen, A.K.; Venema, K.; Reijngoud, D.J.; Bakker, B.M. The Role of Short-Chain Fatty Acids in the Interplay between Diet, Gut Microbiota, and Host Energy Metabolism. *J. Lipid Res.* **2013**, *54*, 2325–2340. [CrossRef]
19. Degnan, P.H.; Barry, N.A.; Mok, K.C.; Taga, M.E.; Goodman, A.L. Human Gut Microbes Use Multiple Transporters to Distinguish Vitamin B_{12} Analogs and Compete in the Gut. *Cell Host Microbe* **2014**, *15*, 47–57. [CrossRef]
20. Sorboni, S.G.; Moghaddam, H.S.; Jafarzadeh-Esfehani, R.; Soleimanpour, S. A Comprehensive Review on the Role of the Gut Microbiome in Human Neurological Disorders. *Clin. Microbiol. Rev.* **2022**, *35*, e0033820. [CrossRef]
21. Liu, J.; Tan, Y.; Cheng, H.; Zhang, D.; Feng, W.; Peng, C. Functions of Gut Microbiota Metabolites, Current Status and Future Perspectives. *Aging Dis.* **2022**, *13*, 1106–1126. [CrossRef]
22. Dmytriv, T.R.; Lushchak, O.; Lushchak, V.I. Glucoraphanin Conversion into Sulforaphane and Related Compounds by Gut Microbiota. *Front. Physiol.* **2025**, *16*, 1497566. [CrossRef] [PubMed]
23. Degnan, P.H.; Taga, M.E.; Goodman, A.L. Vitamin B12 as a Modulator of Gut Microbial Ecology. *Cell Metab.* **2014**, *20*, 769–778. [CrossRef] [PubMed]
24. Stevens, J.F.; Maier, C.S. The Chemistry of Gut Microbial Metabolism of Polyphenols. *Phytochem. Rev.* **2016**, *15*, 425–444. [CrossRef] [PubMed]
25. Gasaly, N.; Gotteland, M. Interference of Dietary Polyphenols with Potentially Toxic Amino Acid Metabolites Derived from the Colonic Microbiota. *Amino Acids* **2022**, *54*, 311–324. [CrossRef]
26. Larabi, A.B.; Masson, H.L.P.; Bäumler, A.J. Bile Acids as Modulators of Gut Microbiota Composition and Function. *Gut Microbes* **2023**, *15*, 2172671. [CrossRef]
27. Chiang, J.Y.; Pathak, P.; Liu, H.; Donepudi, A.; Ferrell, J.; Boehme, S. Intestinal Farnesoid X Receptor and Takeda G Protein Couple Receptor 5 Signaling in Metabolic Regulation. *Dig Dis.* **2017**, *35*, 241–245. [CrossRef]
28. Brito Rodrigues, P.; de Rezende Rodovalho, V.; Sencio, V.; Benech, N.; Creskey, M.; Silva Angulo, F.; Delval, L.; Robil, C.; Gosset, P.; Machelart, A.; et al. Integrative Metagenomics and Metabolomics Reveal Age-Associated Gut Microbiota and Metabolite Alterations in a Hamster Model of COVID-19. *Gut Microbes* **2025**, *17*, 2486511. [CrossRef]
29. Kundu, P.; Lee, H.U.; Garcia-Perez, I.; Tay, E.X.Y.; Kim, H.; Faylon, L.E.; Martin, K.A.; Purbojati, R.; Drautz-Moses, D.I.; Ghosh, S.; et al. Neurogenesis and Prolongevity Signaling in Young Germ-Free Mice Transplanted with the Gut Microbiota of Old Mice. *Sci. Transl. Med.* **2019**, *11*, eaau4760. [CrossRef]
30. Darnaud, M.; De Vadder, F.; Bogeat, P.; Boucinha, L.; Bulteau, A.L.; Bunescu, A.; Couturier, C.; Delgado, A.; Dugua, H.; Elie, C.; et al. A Standardized Gnotobiotic Mouse Model Harboring a Minimal 15-Member Mouse Gut Microbiota Recapitulates SOPF/SPF Phenotypes. *Nat. Commun.* **2021**, *12*, 6686. [CrossRef]

31. Jans, M.; Vereecke, L. A Guide to Germ-Free and Gnotobiotic Mouse Technology to Study Health and Disease. *FEBS J.* **2025**, *292*, 1228–1251. [CrossRef]
32. Haran, J.P.; McCormick, B.A. Aging, Frailty, and the Microbiome-How Dysbiosis Influences Human Aging and Disease. *Gastroenterology* **2021**, *160*, 507–523. [CrossRef]
33. Dmytriv, T.R.; Lushchak, V.I. Gut Microbiome as a Target for Anti-Ageing Interventions. *Subcell. Biochem.* **2024**, *107*, 307–325. [CrossRef]
34. Di Vincenzo, F.; Del Gaudio, A.; Petito, V.; Lopetuso, L.R.; Scaldaferri, F. Gut Microbiota, Intestinal Permeability, and Systemic Inflammation: A Narrative Review. *Intern. Emerg. Med.* **2024**, *19*, 275–293. [CrossRef]
35. Cusumano, G.; Flores, G.A.; Venanzoni, R.; Angelini, P. The Impact of Antibiotic Therapy on Intestinal Microbiota: Dysbiosis Antibiotic Resistance, and Restoration Strategies. *Antibiotics* **2025**, *14*, 371. [CrossRef]
36. Colquhoun, C.; Duncan, M.; Grant, G. Inflammatory Bowel Diseases: Host-Microbial-Environmental Interactions in Dysbiosis. *Diseases* **2020**, *8*, 13. [CrossRef] [PubMed]
37. Cronin, P.; Joyce, S.A.; O'Toole, P.W.; O'Connor, E.M. Dietary Fibre Modulates the Gut Microbiota. *Nutrients* **2021**, *13*, 1655. [CrossRef] [PubMed]
38. Thaiss, C.A.; Zeevi, D.; Levy, M.; Zilberman-Schapira, G.; Suez, J.; Tengeler, A.C.; Abramson, L.; Katz, M.N.; Korem, T.; Zmora, N.; et al. Transkingdom control of microbiota diurnal oscillations promotes metabolic homeostasis. *Cell* **2014**, *159*, 514–529. [CrossRef] [PubMed]
39. Aggarwal, N.; Kitano, S.; Puah, G.R.Y.; Kittelmann, S.; Hwang, I.Y.; Chang, M.W. Microbiome and Human Health: Current Understanding, Engineering, and Enabling Technologies. *Chem. Rev.* **2023**, *123*, 31–72. [CrossRef]
40. Abouelela, M.E.; Helmy, Y.A. Next-Generation Probiotics as Novel Therapeutics for Improving Human Health: Current Trends and Future Perspectives. *Microorganisms* **2024**, *12*, 430. [CrossRef]
41. Lalowski, P.; Zielińska, D. The Most Promising Next-Generation Probiotic Candidates—Impact on Human Health and Potential Application in Food Technology. *Fermentation* **2024**, *10*, 444. [CrossRef]
42. Kumar, A.; Green, K.M.; Rawat, M. A Comprehensive Overview of Postbiotics with a Special Focus on Discovery Techniques and Clinical Applications. *Foods* **2024**, *13*, 2937. [CrossRef] [PubMed]
43. Hijová, E. Postbiotics as Metabolites and Their Biotherapeutic Potential. *Int. J. Mol. Sci.* **2024**, *25*, 5441. [CrossRef] [PubMed]
44. Kasahara, K.; Kerby, R.L.; Zhang, Q.; Pradhan, M.; Mehrabian, M.; Lusis, A.J.; Bergström, G.; Bäckhed, F.; Rey, F.E. Gut Bacterial Metabolism Contributes to Host Global Purine Homeostasis. *Cell Host Microbe* **2023**, *31*, 1038–1053.e10. [CrossRef] [PubMed]
45. Grove, A. The Delicate Balance of Bacterial Purine Homeostasis. *Discov. Bact.* **2025**, *2*, 14. [CrossRef]
46. Morrison, D.J.; Preston, T. Formation of Short Chain Fatty Acids by the Gut Microbiota and Their Impact on Human Metabolism. *Gut Microbes* **2016**, *7*, 189–200. [CrossRef]
47. Pandey, S.; Kawai, T.; Akira, S. Microbial Sensing by Toll-like Receptors and Intracellular Nucleic Acid Sensors. *Cold Spring Harb. Perspect. Biol.* **2014**, *7*, a016246. [CrossRef]
48. Gil, A. Modulation of the Immune Response Mediated by Dietary Nucleotides. *Eur. J. Clin. Nutr.* **2002**, *56* (Suppl. S3), S1–S4. [CrossRef]
49. Swidsinski, A.; Ladhoff, A.; Pernthaler, A.; Swidsinski, S.; Loening-Baucke, V.; Ortner, M.; Weber, J.; Hoffmann, U.; Schreiber, S.; Dietel, M.; et al. Mucosal Flora in Inflammatory Bowel Disease. *Gastroenterology* **2002**, *122*, 44–54. [CrossRef]
50. Wall, R.; Ross, R.P.; Ryan, C.A.; Hussey, S.; Murphy, B.; Fitzgerald, G.F.; Stanton, C. Role of Gut Microbiota in Early Infant Development. *Clin. Med. Pediatr.* **2009**, *3*, 45–54. [CrossRef]
51. Asano, K.; Yoshimura, S.; Nakane, A. Alteration of Intestinal Microbiota in Mice Orally Administered with Salmon Cartilage Proteoglycan, a Prophylactic Agent. *PLoS ONE* **2013**, *8*, e75008. [CrossRef] [PubMed]
52. Yang, B.; Guo, X.; Shi, C.; Liu, G.; Qin, X.; Chen, S.; Gan, L.; Liang, D.; Shao, K.; Xu, R.; et al. Alterations in Purine and Pyrimidine Metabolism Associated with Latent Tuberculosis Infection: Insights from Gut Microbiome and Metabolomics Analyses. *mSystems* **2024**, *9*, e0081224. [CrossRef] [PubMed]
53. Semchyshyn, H.; Lushchak, V.; Storey, K. Possible Reasons for Difference in Sensitivity to Oxygen of Two *Escherichia coli* Strains. *Biochemistry* **2005**, *70*, 424–431. [CrossRef] [PubMed]
54. Vasylkovska, R.; Petriv, N.; Semchyshyn, H. Carbon Sources for Yeast Growth as a Precondition of Hydrogen Peroxide Induced Hormetic Phenotype. *Int. J. Microbiol.* **2015**, *2015*, 697813. [CrossRef]
55. Bayliak, M.; Semchyshyn, H.; Lushchak, V. Possible Accumulation of Non-Active Molecules of Catalase and Superoxide Dismutase in *S. cerevisiae* Cells under Hydrogen Peroxide Induced Stress. *Cent. Eur. J. Biol.* **2007**, *2*, 326–336. [CrossRef]
56. Mayer, E.A. Gut Feelings: The Emerging Biology of Gut-Brain Communication. *Nat. Rev. Neurosci.* **2011**, *12*, 453–466. [CrossRef]
57. Furness, J.B. The Enteric Nervous System and Neurogastroenterology. *Nat. Rev. Gastroenterol. Hepatol.* **2012**, *9*, 286–294. [CrossRef]
58. De Vadder, F.; Kovatcheva-Datchary, P.; Goncalves, D.; Vinera, J.; Zitoun, C.; Duchampt, A.; Bäckhed, F.; Mithieux, G. Microbiota-Generated Metabolites Promote Metabolic Benefits via Gut-Brain Neural Circuits. *Cell* **2014**, *156*, 84–96. [CrossRef]

59. Nwako, J.G.; Patel, S.D.; Roach, T.J.; Gupte, S.R.; Williams, S.G.; Riedman, A.M.; McCauley, H.A. Enteroendocrine Cells Regulate Intestinal Barrier Permeability. *Am. J. Physiol. Cell Physiol.* **2025**, *328*, C1501–C1508. [CrossRef]
60. Jyoti; Dey, P. Mechanisms and Implications of the Gut Microbial Modulation of Intestinal Metabolic Processes. *NPJ Metab. Health Dis.* **2025**, *3*, 24. [CrossRef]
61. Ramadan, Y.N.; Alqifari, S.F.; Alshehri, K.; Alhowiti, A.; Mirghani, H.; Alrasheed, T.; Aljohani, F.; Alghamdi, A.; Hetta, H.F. Microbiome Gut-Brain-Axis: Impact on Brain Development and Mental Health. *Mol. Neurobiol.* **2025**, *62*, 10813–10833. [CrossRef]
62. Kolodziejczyk, A.A.; Zheng, D.; Elinav, E. Diet–Microbiota Interactions and Personalized Nutrition. *Nat. Rev. Microbiol.* **2019**, *17*, 742–753. [CrossRef]
63. Zeevi, D.; Korem, T.; Zmora, N.; Israeli, D.; Rothschild, D.; Weinberger, A.; Ben-Yacov, O.; Lador, D.; Avnit-Sagi, T.; Lotan-Pompan, M.; et al. Personalized Nutrition by Prediction of Glycemic Responses. *Cell* **2015**, *163*, 1079–1094. [CrossRef]

biomolecules

MDPI

Review

Non-Pancreatic Digestive Enzymes

Drucy Borowitz

Department of Pediatrics, Jacobs School of Medicine and Biomedical Sciences, University at Buffalo, Buffalo, NY 14203, USA; borowitz@buffalo.edu; Tel.: +1-(716)-323-0000; Fax: +1-(716)-323-0290

Abstract: Although the pancreas is the organ that produces the most critical digestive enzymes, there are other important contributors to the cleavage of food into absorbable units. Pre-pancreatic digestion of carbohydrates occurs through the action of salivary amylase. Pre-pancreatic digestion of fats is mediated by lingual and gastric lipases, and their action may be important as a signal for coordinated digestion. Pepsin, which is present in the stomach, initiates the digestion of dietary proteins into peptides and amplifies distal proteolysis. The major post-pancreatic intestinal carbohydrate-digesting enzymes are sucrase-isomaltase, maltase-glucoamylase and lactase-phlorizin hydrolase. There are no post-pancreatic mucosal enzymes that act on dietary triglycerides; however, the complete digestion of phospholipids depends on several brush border phospholipases. Intestinal processing is an important contributor to digestion of proteins, although mucosal proteases may serve as signaling proteins rather than as primary adjuncts to dietary protein digestion and absorption. This review describes the role of these non-pancreatic digestive enzymes in supporting nutritional health.

Keywords: amylase; lipase; protease; gastrin; maltase; sucrase; lactase; peptidases; malabsorption

1. Overview

Digestion involves the enzymatic breakdown of dietary intake into components that can be absorbed and used for growth. The pancreas is the organ that produces the most critical enzymes, and individuals with exocrine pancreatic insufficiency have malabsorption because of loss of pancreatic secretions. These secretions neutralize gastric acid and provide the elements needed for the assimilation of dietary components. To prevent malnutrition, exocrine pancreatic insufficiency is managed by the provision of exogenous pancreatic enzyme replacement therapy. However, there are other important contributors to the cleavage of food into absorbable units. These other enzymes interact with substrates in the mouth and stomach prior to pancreatic secretion and along the intestinal brush border together with and following the action of pancreatic enzymes. Intestinal bacteria enzymes can also digest components of the diet and, in particular, may support colonic health [1]; however, this review will be limited to human intestinal enzymes. Perception of fats in the mouth increases pancreatic secretion and gastric digestion releases amino acids and free fatty acids. These control the gastric emptying rate and stimulate pancreatic enzyme secretion [2]. Post pancreatic digestion enables the process of digestion to be completed. This review will describe these human non-pancreatic digestive enzymes (Table 1), their actions, and how this intricate orchestration of the processing of food supports nutritional health.

Table 1. Summary of non-pancreatic digestive enzymes.

	Carbohydrates	Fats	Proteins
Pre-pancreatic	• Salivary amylase	• Lingual lipase • Gastric lipase	• Pepsin
Post pancreatic	• Sucrase-isomaltase • Maltase-glucoamylase • Trehelase • Lactase	• Phospholipase B1 • Alkaline sphingomyelinase Neutral ceramidase	• Aminopeptidase N • Carboxypeptidase P • Dipeptidyl aminopeptidase IV • Glutamyl aminopeptidase • Angiotensin-converting enzyme

2. Pre-Pancreatic Digestion

2.1. Pre-Pancreatic Digestion of Carbohydrates: Salivary Amylase

The mechanical digestion of food commences with the act of chewing, but enzymatic digestion also begins in the mouth. Saliva is secreted to wet food and enable it to be swallowed. Amylolysis by salivary amylase leads to a rapid decrease in glucose-polymer chain length and viscosity, facilitating the swallowing of food [3]. The products of α-amylase are dextrin fragments, maltose and isomaltose. Salivary and pancreatic amylases are similar but are encoded by different genes (AMY1 and AMY2, respectively). These amylases show different levels of activity against starches of various origins [4,5]. AMY1 is present at high levels in the parotid and lingual salivary glands. Human salivary amylase activity may have evolved as agriculture and harvesting of starches from grains and tubers supplemented hunting in our diets. The levels of salivary amylase in humans are by far the highest among primates [6]. Obligate meat-eaters, such as cats, never have salivary amylase.

Salivary amylase may have functions not directly related to its actions on the digestion of food. Data in rats shows that the AMY1 gene is expressed in taste glands, although at low levels, suggesting that it may have a role in food perception exclusive of the action of amylase on the texture of starches [7]. Others have suggested that humans can detect oligosaccharides independent of the sense of sweetness provided by glucose [8]. Salivary amylase also binds strongly to some oral streptococci, a function that may have a role in cariogenesis [9]. Although not related to its enzymatic activity, quantitation of salivary amylase has been used as an objective measure of behavioral response to stress as it reflects autonomic nervous system activity [10,11].

Salivary amylase is present and active in premature infants and levels appear to be stable across all age groups [12,13]. Although its amylolytic activity is small compared to that of pancreatic amylase, it may serve as compensation for lack of pancreatic amylase in patients with exocrine pancreatic insufficiency [14,15]. Encouraging extended mastication does not appear to be of benefit in increasing the amount of salivary amylase. Although studies vary in their conclusions, a systematic review of the literature concluded that "chewing in healthy people has minimum effect on the expression and activities of salivary proteins" [16].

2.2. Pre-Pancreatic Digestion of Fats

Most dietary fats are present as triglycerides found in meats, fish, dairy, oils, nuts and seeds. Pre-pancreatic enzymes that digest triglycerides are secreted in the mouth and stomach. Both lipases are active in an acid environment. Lingual and gastric lipases are present in the fetus from 26 weeks of gestation onwards. The activity of pre-pancreatic lipases may be a compensatory mechanism for digestion in patients with exocrine pancreatic insufficiency [17,18]. The coefficient of fat absorption in patients with exocrine pancreatic insufficiency who do not secrete pancreatic lipase is not zero, likely reflecting the activity of pre-pancreatic lipases [19–22].

2.2.1. Lingual Lipase

Release of lingual lipase in the mouth is signaled by the autonomic nervous system following the start of a meal. The von Ebner's glands on the surface of the tongue secrete lingual lipase in areas near fat taste receptors. A study in human neonates demonstrated a pH optimum of 3.5–6 and suggested that lingual lipase remains active in the fed stomach [23]. The action of lingual lipase releases non-esterified fatty acids (NEFA) from dietary fats. NEFA are signaling molecules. Although lingual lipase in rodents plays a key role in oral detection of fats, this effect does not seem to be as strong in humans [24]. Nonetheless, NEFA are released in humans in concentrations sufficient to depolarize taste receptor cells [25].

2.2.2. Gastric Lipase

The proportion of lingual to gastric lipase varies among species. In the baboon and human, gastric lipase is present in much greater amounts than lingual lipase [26]. Lipolysis is greatest in the fundus of the stomach and is lowest in the antrum [27]. The biochemical properties of gastric lipase are the subject of an excellent review by Sams et al. [28]. Gastric lipase is stable and active in the pH range 2 to 7 with an optimal cleavage of fats at pH 4–5.4, the pH of the fed stomach. Under those conditions, 10–25% of acyl chains are released from dietary triglycerides [18,29]. Gastric lipase is resistant to degradation by pepsin. Like pancreatic lipase, the active site is controlled by a lid domain. Human gastric lipase is highly interactive with tri- and di-glycerides forming emulsions and has less activity on monoglycerides [30]. Species differences occur however gastric lipase cannot cleave phospholipids and cholesterol esters. Gastric lipase can hydrolyze all three ester bonds of triglycerides but preferentially cleaves the ester bond at the sn-3 position.

Another unique quality of human gastric lipase is that it changes the surface tension of droplets to allow bile salts to penetrate the phospholipid layers covering triglycerides. Unlike pancreatic lipase, gastric lipase is active in the presence of bile salts without needing a co-factor [31]. Human pancreatic lipase requires co-lipase, which stabilizes its active conformation, anchors it to the water–lipid interface, and displaces bile salts. Gastric lipase can initiate hydrolysis of milk triacylglycerol and thus is important for digestion of milk fat. Another pre-pancreatic lipase, bile-salt-stimulated lipase, is found in human milk. Activated bile salt-stimulated lipase cannot hydrolyze native milk fat globules. It is dependent on partial hydrolysis by gastric lipase, which triggers bile salt-stimulated lipase and pancreatic lipase activity [32].

Gastric lipolysis may be important as a signal for coordinated digestion as well as appetite and growth. Fatty acids, the products of gastric lipolysis, trigger the activity of pancreatic lipase [33]. The critical importance of gastric lipase is demonstrated in subjects who have undergone gastrectomy, whose regulation of pancreatic secretion becomes abnormal after gastric surgery, when gastric digestion and emptying are altered [34,35]. In addition, ghrelin is a gut hormone that aids the process of food uptake and growth hormone release and requires octanylation to be active [36]. Gastric lipase releases short and medium chain fatty acids such as octanoic acid that are specifically found at the sn-3 position of dairy products and various oils. It has been hypothesized that role of gastric lipase in releasing octanoic acid for ghrelin acylation is important for growth [28].

2.3. *Pre-Pancreatic Digestion of Proteins: Pepsin*

Dietary proteins are found in a range of foods including meat and fish as well as beans and nuts. Pepsin initiates the digestion of dietary proteins into peptides and amplifies distal proteolysis. In response to the hormone gastrin and to vagal stimulation, chief cells in the stomach secrete pepsinogen, an inactive zymogen that is converted to pepsin. Gastric acid is required to convert pepsinogen into pepsin. Pepsin is active at pH = 2, and transforms to an inactive conformation at pH 4.0–6.5 [37]. The digestive specificity of pepsin is broad, but the presence of phenylalanine or leucine increases the cleavage probability of porcine pepsin to greater than 40% and the presence of histidine, lysine, arginine or proline prohibits cleavage when found at the P1 position [38]. The strongest amino acid stimulants of Ca^{2+}-sensing receptors, phenylalanine and tryptophan, are released by the action of pepsin [39]. Ca^{2+}-sensing receptors control the release of gastrin and CCK from gastric acid-secreting parietal cells and neuroendocrine cells, thus increasing proteolysis [39,40]. Thus, pepsin both digests proteins directly and generates signals to aid digestion.

3. Post-Pancreatic Digestion

The intestinal brush border has microvilli that are the site for a wide range of digestive enzymes. These enzymes may be anchored to the brush border, and they can become incorporated into vesicles that are shed near the apices of microvilli. This helps nutrient hydrolysis to occur close to the absorptive surface [41].

3.1. Post-Pancreatic Digestion of Carbohydrates

Pancreatic α-amylase, which is more potent than salivary amylase, continues to digest the large oligosaccharides (α-dextrins) converting them to disaccharides (maltose), trisaccharides (maltotriose), and oligosaccharides called limit dextrins which have four to nine glucosyl residues and an isomaltose branch. Alpha-amylase messenger RNA is found at high levels in the duodenum. This duodenal amylase appears to have a role in maintaining intestinal epithelial integrity and at high concentrations, inhibits glucose uptake by the brush border Na+/glucose cotransporter 1 (SGLT1), thus moderating high blood sugar swings from high dietary glucose loads [42]. Further digestion of the products released by amylase is mediated by other enzymes located on the intestinal epithelium. Trehalose, lactose and sucrose are not acted upon by amylase but there are specific enzymes in the small intestine to aid with digestion of these dietary components.

The major intestinal carbohydrate-digesting enzymes, including sucrase-isomaltase (SI), maltase-glucoamylase (MGAM) and lactase-phlorizin hydrolase are membrane bound and are not secreted. They require immediate contact with their substrates. As demonstrated by Amiri et al., respectively, they constitute 8.2%, 2.7% and 1.4% of total brush border membrane protein and show optimal activity at pH 6, with over 50% residual activity between pH 5 to pH 7 [43]. The substrates, their corresponding enzymes, and the subunits to which they are digested are in Table 2.

Table 2. Digestion of carbohydrates.

	Enzyme	Carbohydrate Subunits	Bond Cleaved	Examples of Foods Providing Substrate
Products of amylase digestion:				
Isomaltose	Isomaltase	Glucose+glucose	α1,6	Wheat, cornmeal
Maltose	Maltase II	Glucose+glucose	Primarily α1,4 but also α-1,2, α-1,3 and α-1,6	
Limit dextrin	Glucoamylase	Glucose polymer with isomaltose branch	α1,4; cleaves from the non-reducing end	

Table 2. *Cont.*

	Enzyme	Carbohydrate Subunits	Bond Cleaved	Examples of Foods Providing Substrate
Dietary disaccharides:				
Sucrose	Sucrase	Glucose+fructose	α1,2	Some fruits and vegetables
Trehalose	Trehelase	Glucose+glucose	α1,1	Mushrooms, baker's yeast
Lactose	Lactase	Glucose+galactose	β1,4	Dairy products

3.1.1. Alpha Glucosidases

The name alpha-glucosidases describes four maltases, enzymes that convert linear starch into simple sugars. Two of these maltase activities are associated with sucrase-isomaltase (maltase Ib, maltase Ia). The other two are named maltase-glucoamylase (maltases II and III). The sucrase–isomaltase and maltase-glucoamylase complexes have similar structures and exhibit much sequence homogeneity. The protein is attached to the luminal membrane by a domain near the N-terminus. Complete digestion of α-linked glucose polymers to absorbable glucose takes place through the action of the 4 mucosal α-glucosidases in the small intestine.

Sucrase-Isomaltase

The sucrase-isomaltase (SI) complex has two subunits that digest different substrates. It is synthesized as a single polypeptide chain which is divided into two subunits, one of which remains attached to the small intestinal membrane. Eighty percent of the maltose that we eat is digested by this complex. Almost all the isomaltose formed by glucoamylase from limit dextrins is hydrolyzed by the SI complex. Levels of SI are highest in the jejunum and are lower in the duodenum and ileum [44].

Congenital sucrase-isomaltase deficiency (CSID), now called genetic SI deficiency, is the inability to digest sucrose and starch due to mutations in the SI gene that lead to no function. Maldigested sugars accumulate in the bowel leading to osmotic diarrhea, increased motility, and increased bacterial fermentation causing intestinal gas and bloating. Although classically it has been described as a disorder identified in infancy when solid foods and juices are added to the diet, it may present later in life in patients with symptoms similar to those seen in irritable bowel syndrome (IBS) [45]. These individuals have genetic variants that are milder than those seen in CSID. A diet low in carbohydrates that are poorly absorbed in the intestine is commonly used to control symptoms in patients with IBS, however it does not restrict sucrose (FODMAP diet; fermentable oligosaccharides, disaccharides, monosaccharides, and polyols). In one study, 50% of patients with diminished response to a low FODMAP diet were carriers of an SI variant that decreases SI enzymatic activity [46].

Identification of novel mutations has led to the categorization of genetic SI deficiency into three phenotypes based on the activity of the variants on protein trafficking, enzyme function and lipid raft association. Levels of enzyme activity can vary over a large range depending on the mutation [47]. Genetic SI deficiency occurs in 0.05–0.2% of individuals of European descent and approximately 5–10% among indigenous Greenlanders [48]. Inheritance is autosomal recessive and carrier frequencies have been reported as near 2% in non-Hispanic Whites, around 0.5% in Hispanic Whites and less than 0.05% in African Americans [48,49].

Maltase-Glucoamylase

The maltase-glucoamylase complex is like the SI complex but it has only one polypeptide chain. Maltase digests α-1,2- and α-1,3-disaccharides better than the other α-glucosidases. In addition, it can hydrolyze α-1,4 and α-1,6 linkages [50]. Amylase cleaves bonds within the interior of starch molecules. In contrast, glucoamylase is an exoglycosidase that begins enzymatic cleavage at the nonreducing end of a polysaccharide or limit dextrin, releasing glucose monosaccharides. It will digest a limit dextrin down to isomaltose, which can then be hydrolyzed by the sucrase–isomaltase complex. Maltase hydrolyzes small-sized malto-oligosaccharides to glucose; glucoamylase preferentially hydrolyzes longer malto-oligosaccharides. In humans, maltase activity increases along the intestine and reaches its highest activity in the distal ileum [44].

Although less common than SI deficiency, maltase-glucoamylase deficiency was documented by intestinal biopsy in 25% of 203 children with chronic abdominal pain [51]. These individuals had a normal gross appearance to the bowel, which is important because damage to the intestinal mucosa can cause secondary dissacharidase deficiencies. In a study by Viswanathan and Rao, 120 adult patients with unexplained gastrointestinal (GI) symptoms, 0.8% were maltase deficient, 9.2% of patients had deficiencies in all 4 α-glucosidases, and 0.8% had combined sucrase, maltase and isomaltase deficiency [52].

Trehalase

Trehalase is a brush border α-glycosidase. Trehalose is synthesized in many organisms where it may serve as a source of energy and carbon, but humans do not generate trehalose. Rather, the intestinal enzyme is present to digest this sugar, which is found primarily in mushrooms and baker's yeast. A study in rats found trehalase mRNA expression to be highest in the duodenum, decreasing towards the distal ileum [53]. Isolated trehalase deficiency occurs in at least 8% of the Greenlanders but is extremely rare in intestinal biopsies of other populations [54,55].

3.1.2. Lactase-Phlorizin Hydrolase

Lactose is not metabolized by amylase (which can only cleave α-glycolytic bonds) and thus enters the small intestine intact, where it is degraded by lactase-phlorizin hydrolase, commonly called lactase. Lactase is also termed β-glycosidase because it is the only brush border enzyme that hydrolyzes β- glycosidic bonds. Lactase acts on lactose and glycolipids to form glucose and galactose. Levels were found to be highest in human jejunum with decreasing activity towards the proximal and distal ends of the intestine [44].

The inability to digest lactose may be congenital, developmental, primary, or secondary. Congenital lactose intolerance is rare. It presents with severe diarrhea in the first days of life because of the complete inability to digest milk, the primary food source for infants [56,57]. Lactase activity increases during fetal development. Between 26 and 34 weeks of gestation, lactase activity is approximately one third of that found in full term babies. From 35 to 38 weeks of gestation, lactase activity increases to about 70% of term infants [58]. Thus, prematurely born infants have developmental lactose intolerance. Primary lactose intolerance results from decreased synthesis of lactase, either at the transcriptional or translational level [59]. As per the American Academy of Pediatrics, "Approximately 70% of the world's population has primary lactase deficiency. . .In populations with a predominance of dairy foods in the diet, particularly northern European people, as few as 2% of the population has primary lactase deficiency. In contrast, the prevalence of primary lactase deficiency is 50% to 80% in Hispanic people, 60% to 80% in black and Ashkenazi Jewish people, and almost 100% in Asian and American Indian people. . . The age of onset and its prevalence differ among various populations. Approximately 20% of

Hispanic, Asian, and black children younger than 5 years of age have evidence of lactase deficiency and lactose malabsorption, whereas white children typically do not develop symptoms of lactose intolerance until after 4 or 5 years of age." [60]. Adults with unexplained GI symptoms may have lactose intolerance, either alone or in combination with deficiency of all four disaccharidases ("pan-disaccharidase deficiency") [52]. Secondary lactase deficiency results when there is small intestinal injury with blunting of the intestinal villi and subsequent loss of lactase at their tips. Severe malnutrition, after enteric infections or with celiac disease, Crohn disease or other immune-related injury can lead to secondary lactase deficiency which improves when the intestinal mucosa has healed.

3.2. Post-Pancreatic Digestion of Fats

Although there are also cytoplasmic lipases, our focus is on brush border enzymes. The majority of lipid digestion and absorption occurs in the duodenum, primarily through the action of gastric and pancreatic lipases. The major dietary lipid class, triglycerides, are rapidly hydrolyzed by the combined action of gastric and pancreatic lipases in the stomach and upper small intestine. There are no post-pancreatic mucosal enzymes that act on triglycerides. However, the digestion of phospholipids is more extended. Phospholipids are found in foods such as eggs, organ meats, fish, shellfish, cereal grains, and oilseeds, as well as plant sources like soybeans, corn, and cruciferous vegetables. There are two types of phospholipids:

(1) Glycerophospholipids are key components of biological membranes. They have a hydrophobic tail that can become embedded in the cell's membrane, connected by a glycerophosphate component to a hydrophilic head, which can be moieties such as choline, ethanolamine, or serine, among others (Figure 1) [61]. The pancreas assists in digestion of glycerophospholipids by secreting phospholipases A_1 and A_2 (PLA_1 and PLA_2, respectively) but studies in the PLA_2 knock-out mouse suggest that distal, mucosal phospholipases play a major role in digestion and absorption of this class of lipids [62].

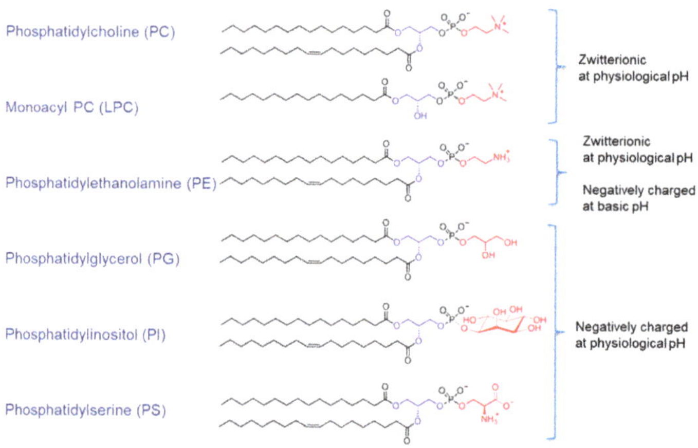

Figure 1. Structure of phospholipids. https://www.phospholipid-research-center.com/phospholipid/types/ (accessed on 29 August 2025). Reproduced with permission from van Hoogevest P et al., Eur. J. Lipid Sci. Technol. [61], Figure 2, published by Wiley-VCH GmbH 2021.

(2) Sphingophospholipids, such sphingomyelin and ceramide-1-phosphate, are important constituents of nervous system tissues. There are no pancreatic enzymes that digest sphingohospholipids.

In addition to glycerophospholipases and sphingophospholipases, there is a brush border-associated retinyl esterase that hydrolyzes sources of pre-formed vitamin A, mostly found in animal products such as liver, eggs, and dairy products.

3.2.1. Glycerophospholipids

Phospholipase B_1 (PLB): Pancreatic phospholipases A_1 and A_2 hydrolyze the ester bond at the sn-1 and sn-2 positions of glycerophospholipids, respectively. Mucosal PLB is a membrane-bound, calcium-independent enzyme with broad substrate specificity and a combination of both PLA_1 and PLA_2 activities, although it preferentially cleaves sn-2 ester bonds over sn-1 bonds. PLB plays an important role in digestion in the jejunum and ileum [63]. In addition, it may act as a retinyl esterase. Dietary retinol (Vitamin A) can be directly absorbed. Retinyl esters are fatty acids that are esterified to retinol. These cannot be absorbed until hydrolysis yields free retinol. The pancreas secretes retinyl esterases that have the greatest activity for esters whose fatty acyl portions are $\leq$10 carbons in length. Brush border retinyl esterase is active in the presence of bile salts and has its greatest activity toward the long-chain retinyl esters found in the diet [64]. The encoded protein is highly conserved.

3.2.2. Sphingophospholipids

Alkaline Sphingomyelinase (Alk-SMase)

There are no pancreatic enzymes that hydrolyze sphingomyelin. Alk-SMase, the key enzyme responsible for SM hydrolysis, is present on the surface of the microvilli. It is released in active form into the intestinal lumen by bile salt and trypsin digestion [65]. It is protease resistant and acts in an alkaline environment. It is expressed neonatally. Decreased alk-SMase activity has been found in colon cancer and colitis [66,67].

Neutral Ceramidase

Ceramides are two-chained sphingolipids that are part of the structure of sphingomyelin. Thus, neutral ceramidase works in concert with Alk-SMase to hydrolyze ceramide to sphingosine and free fatty acids (Figure 2). Ceramide plays an important role in regulating cell proliferation and cell death and because of its role in complementing Alk-SMase it has also been implicated in the development of colon cancer. Data in animals suggests that neutral ceramidase activity is low in the proximal duodenum, increases to a plateau in the proximal jejunum and declines in the distal part of ileum. Some activity is also detectable in the colon [68]. Like alk-SMase it is stimulated by bile salts, is protease resistant, and is expressed neonatally.

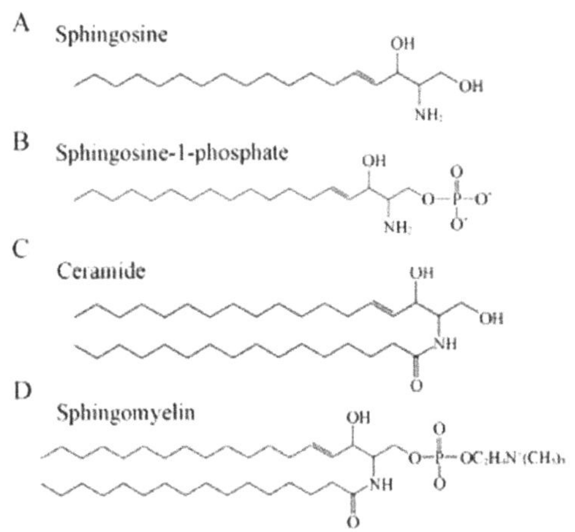

Figure 2. Structure of sphingophospholipids. https://www.creative-proteomics.com/blog/index.php/what-is-sphingomyelin/ (accessed on 29 August 2025). Reproduced with permission from Creative Proteomics 2025.

3.3. Post-Pancreatic Digestion of Proteins

As with phospholipases, there are cytoplasmic peptidases, but our focus is on brush border enzymes. Brush border proteases are more correctly termed peptidases since they hydrolyze specific peptide bonds. Brush border enterokinase converts the proenzyme trypsinogen, which is secreted by the pancreas, to yield active trypsin. Enterokinase is a brush border peptidase but is not technically part of post-pancreatic digestion since it activates pancreatic protease. The activity of many of these brush border peptidases can be detected by 10–16 weeks of gestation and increases during fetal life. Many serve as signaling proteins rather than as primary adjuncts to dietary protein digestion and absorption. However, a study in rats demonstrated that 30% of ingested proteins were absorbed from isolated small intestinal loops following pancreatic duct ligation: larger amounts were absorbed if pepsin or pepsin plus acid were added to the lumen [69]. Thus, intestinal processing is an important contributor to digestion of proteins.

With the exception of pancreatic carboxypeptidases, pancreatic proteolytic enzymes are endopeptidases, cleaving peptide bonds within the polypeptide chain. Exopeptidases remove amino acids from the terminal ends of oligopeptides. Many brush border exopeptidases have been identified: what follows is a representative list.

3.3.1. Aminopeptidase N

(APN/CD13) Aminopeptidase N (APN) is an N-terminal exopeptidase and is a membrane protein found in large amounts in the microvilli. In a pig model it represents more than half of the peptidase activity in the jejunum [70]. This zinc-dependent enzyme has three main functions: cleaving amino acids from the N-terminus of peptides, endocytosis (including of cholesterol), and signaling [71]. The digestive function of APN (and other brush border peptidases) is to convert proteins into oligopeptides and free amino acids that can subsequently be absorbed into the enterocyte. In addition, it cleaves a wide variety of metabolically active substrates such as enkephalins, kinins, and angiotensins, among others [71]. Enzyme activity increases in a proximal to distal manner in the small intestine and reaches maximum values in distal ileum [44].

3.3.2. Dipeptidyl Aminopeptidase IV

(Dipeptidyl peptidase-4; DPP-4; CD26) enzymatically digests peptides, particularly those with alanine, proline, or serine residues in the penultimate position from the N-terminus. More than as an adjunct to incorporation of dietary protein for anabolism, aids in regulating metabolism and immune responses. DPP-4 helps degrade incretin hormones like glucagon-like peptide-1 (GLP-1) and glucose-dependent insulinotropic polypeptide (GIP), which are important for regulating blood sugar levels. Increased levels of GLP-1 lead to increased insulin and decreased glucagon production, which lowers blood glucose. DPP-4 inhibitors, also known as gliptins, are medications used to treat type 2 diabetes [72]. By blocking the action of DPP-4, levels of incretins increase, leading to lower blood glucose. Maximum activity is in the distal ileum [44].

3.3.3. Glutamyl Aminopeptidase (ENPEP; Aminopeptidase A)

Glutamyl aminopeptidase is a zinc-dependent aminopeptidase that cleaves glutamatic and aspartatic acid residues from the N-terminus of polypeptides. In humans, ENPEP expression is highest in the terminal ileum [44,73].

3.3.4. Carboxypeptidases

Carboxypeptidases remove amino acids from the C-terminal end of oligopeptides, rather than the N-terminal end. They are ubiquitous in the body and serve a wide range of functions in addition to digestion of dietary proteins. They also contribute to the regulation of blood fibrinolysis and the maturation of neuropeptides and hormones, among others. Pancreatic carboxypeptidases A1, A2 and B release a wide range of amino acids with a preference for those with small aliphatic and bulky aromatic side chains. However, pancreatic enzymes cannot release acidic amino acids from the C-terminus of polypeptides, such as glutamic or aspartic acid, even though these are among the most abundant amino acids in dietary proteins. Brush border carboxypeptidase O has this capacity and together with the action of the pancreatic carboxypeptidases, they can digest most proteins [74]. Carboxypeptidase activity is distributed throughout the small intestine [44].

3.3.5. Angiotensin-Converting Enzymes (ACE and ACE2)

ACE is also a carboxypeptidase. It is present in the jejunal brush border where it aids in nitrogen absorption but likely plays a larger role in regulating local levels of signaling peptides. ACEs are present in many cells in the body. Vascular ACE is associated with the renin–angiotensin system, regulating peripheral blood pressure by converting angiotensin I to angiotensin II, inactivating bradykinin, and hydrolyzing enkephalins and other neuroactive peptides [75]. Epithelial intestinal ACE may act similar to its endothelial counterpart by controlling levels of endogenous bioactive peptides known to affect jejunal flows of sodium, chloride, bicarbonate and water. The GI renin–angiotensin system and its interactions with the GI microbiome are complex [76]. ACE2 has 40% overall identity to ACE and stabilizes neutral amino acid transporters. When not present, intestinal uptake of dietary amino acids such as tryptophan, is affected [77]. ACE2 is a SARS-CoV-2 receptor, and by down-regulating intestinal ACE2, SARS-CoV-2 could promote leaky gut syndrome, modify the gut microbiome and enhance systemic inflammation. As with DPP-4, the action of GI ACEs are less as true dietary digestive enzymes and more as regulators of gut homeostasis.

3.3.6. A Few Endopeptidases Are Also Found in the Small Intestine

The most prevalent GI endopeptidase is neprilysin, which has broad substrate specificity. Intestinal neprilysin plays a role in regulating glucose homeostasis by inactivating incretins, including GLP-1. Meprin A is an endopeptidase that hydrolyzes peptides and insulin beta-chains [41].

4. Clinical Relevance

Digestion is a complex process and fortunately one with some redundancy. The most notable example is the compensation seen in patients with exocrine pancreatic insufficiency (EPI). EPI is seen in patients with cystic fibrosis, Schwachman-Diamond Syndrome, chronic pancreatitis, and pancreatic cancer. Although untreated EPI can lead to life-threatening malnutrition, it is also notable that malabsorption of fat and protein is diminished but not totally absent even in the absence of treatment, likely owing to the combined effects of pre- and post-pancreatic digestive enzymes. Clinicians often prescribe acid blocking medicines along with pancreatic enzyme replacement therapy for patients with EPI to overcome the inactivation of pancreatic lipase and protease by gastric acid. The natural stability of lingual and gastric lipase and of pepsin suggests that this strategy is not necessary to enhance pre-pancreatic digestion [23,37].

If there is structural or functional damage to the intestinal lining, brush border enzymes are reduced. This can be seen with severe diarrheal disease, celiac disease, small intestinal involvement in Crohn Disease or graft-versus-host intestinal damage and leads to nutrient malabsorption and consequent GI symptoms. The loss of brush border enzymes can lead to significant disruption of digestion. A congenital absence of brush border enterokinase leads to symptoms of protein malabsorption including failure to thrive and edema [78]. Infants with genetic sucrase-isomaltase deficiency present with osmotic diarrhea, distention and abdominal pain early in life, but partial expression of less severe genetic variants may not have overt symptoms until later in life [45]. As noted above, these genetic variants can be a cause of some symptoms of irritable bowel syndrome [46]. When the brush border is damaged there can also be deficient enteric hormone signaling of pancreatic secretion (secondary pancreatic insufficiency), so the presence of malabsorption symptoms seen with intestinal mucosal damage is multifactorial. Small intestinal bacterial overgrowth (SIBO) can be a cause of mucosal damage or can be a result of disruption of enteric integrity from any of the diseases noted above. SIBO can also be a consequence of dysmotility and is seen in patients with cystic fibrosis [79]. SIBO interferes with brush border enzyme activities through the action of bacterial proteases on the mucosa and leads to symptoms of malabsorption with similar pathophysiology to that seen with overt mucosal damage.

Short bowel syndrome (SBS), such as from congenital malformations or significant small intestine surgical resection, is the leading cause of intestinal failure in children [80] . The functional definition of SBS considers "the critical loss of the length of gut mass or its function below a minimum required for the adequate absorption of enteral nutrients and fluids to provide essential nourishment for the growth and the maintenance of life" [81]. Intestinal rehabilitation programs are designed to promote intestinal absorption. Controlling symptoms of carbohydrate malabsorption such as bloating or/and diarrhea resulting from loss of sucrase-isomaltase and lactase activity is important. Although not an explicit focus of current management, attention to the loss of non-pancreatic proteolytic enzymes, deserves attention. For example, loss of jejunum and thus activity of phospholipase B_1 and neutral ceramidase can contribute to phospholipid malabsorption as well as that of other lipids and retinyl esters. Partial loss of the activity of DPP4 from distal ileal resection has the potential to lead to metabolic and immune system dysfunction. GLP-2 improves nutrient absorption and gut adaptation in rodents or humans with short bowel syndrome and has been proposed as a potential treatment [82]. Whether GLP-2 acts by increasing the activity of post-pancreatic digestive enzymes is an area for future exploration.

5. Conclusions

Although the pancreas is the organ that produces the most critical enzymes, there are other important contributors to the cleavage of food into absorbable units. These other enzymes interact with substrates in the mouth and stomach prior to pancreatic secretion and along the intestinal brush border in combination with and following the action of pancreatic enzymes. Awareness of these other components of digestion can be of help when assessing patients with GI symptoms such as bloating, abdominal pain or diarrhea and may be important when considering metabolic or nutritional deficiencies in patients with underlying intestinal disorders. Additionally, the efficacy of pancreatic enzyme replacement therapy cannot be assessed by improvement in symptoms alone. For example, ongoing bloating or abdominal pain may be caused by poor activity of post-pancreatic amylolytic enzymes. The aim of this review is to remind clinicians that digestion and absorption is a complex process and although important, it is not limited to the action of pancreatic enzymes.

Funding: This research received no external funding.

Institutional Review Board Statement: Not applicable.

Informed Consent Statement: Not applicable.

Data Availability Statement: No new data were created or analyzed in this study. Data sharing is not applicable to this article.

Conflicts of Interest: The author declares no conflict of interest.

References

1. Slavin, J. Fiber and prebiotics: Mechanisms and health benefits. *Nutrients* **2013**, *5*, 1417–1435. [CrossRef]
2. Fieker, A.; Philpott, J.; Armand, M. Enzyme replacement therapy for pancreatic insufficiency: Present and future. *Clin. Exp. Gastroenterol.* **2011**, *4*, 55–73. [CrossRef] [PubMed]
3. Evans, I.D.; Haisman, D.R.; Elson, E.L.; Pasternak, C.; McConnaughey, W.B. The effect of salivary amylase on the viscosity behaviour of gelatinised starch suspensions and the mechanical properties of gelatinised starch granules. *J. Sci. Food Agric.* **1986**, *37*, 573–590. [CrossRef]
4. Meites, S.; Rogols, S. Serum amylases, isoenzymes, and pancreatitis. I. Effect of substrate variation. *Clin. Chem.* **1968**, *14*, 1176–1184. [CrossRef]
5. Hall, F.F.; Ratliff, C.R.; Hayakawa, T.; Culp, T.W.; Hightower, N.C. Substrate differentiation of human pancreatic and salivary alpha-amylases. *Am. J. Dig. Dis.* **1970**, *15*, 1031–1038. [CrossRef] [PubMed]
6. Perry, G.H.; Dominy, N.J.; Claw, K.G.; Lee, A.S.; Fiegler, H.; Redon, R.; Werner, J.; Villanea, F.A.; Mountain, J.L.; Misra, R.; et al. Diet and the evolution of human amylase gene copy number variation. *Nat. Genet.* **2007**, *39*, 1256–1260. [CrossRef] [PubMed]
7. Merigo, F.; Benati, D.; Cecchini, M.P.; Cristofoletti, M.; Osculati, F.; Sbarbati, A. Amylase expression in taste receptor cells of rat circumvallate papillae. *Cell Tissue Res.* **2009**, *336*, 411–421. [CrossRef]
8. Lapis, T.J.; Penner, M.H.; Lim, J. Evidence that humans can taste glucose polymers. *Chem. Senses* **2014**, *39*, 737–747. [CrossRef]
9. Scannapieco, F.A.; Torres, G.; Levine, M.J. Salivary alpha-amylase: Role in dental plaque and caries formation. *Crit. Rev. Oral. Biol. Med.* **1993**, *4*, 301–307. [CrossRef]
10. Nater, U.M.; Rohleder, N. Salivary alpha-amylase as a non-invasive biomarker for the sympathetic nervous system: Current state of research. *Psychoneuroendocrinology* **2009**, *34*, 486–496. [CrossRef]
11. Ali, N.; Nater, U.M. Salivary Alpha-Amylase as a Biomarker of Stress in Behavioral Medicine. *Int. J. Behav. Med.* **2020**, *27*, 337–342. [CrossRef]
12. Murray, R.D.; Kerzner, B.; Sloan, H.R.; McClung, H.J.; Gilbert, M.; Ailabouni, A. The contribution of salivary amylase to glucose polymer hydrolysis in premature infants. *Pediatr. Res.* **1986**, *20*, 186–191. [CrossRef]
13. Aguirre, A.; Levine, M.J.; Cohen, R.E.; Tabak, L.A. Immunochemical quantitation of alpha-amylase and secretory IgA in parotid saliva from people of various ages. *Arch. Oral. Biol.* **1987**, *32*, 297–301. [CrossRef]
14. Skude, G.; Kollberg, H. Serum isoamylases in cystic fibrosis. *Acta Paediatr. Scand.* **1976**, *65*, 145–149. [CrossRef]
15. Townes, P.L.; Moore, W.D.; White, M.R. Amylase polymorphism: Studies of sera and duodenal aspirates in normal individuals and in cystic fibrosis. *Am. J. Hum. Genet.* **1976**, *28*, 378–389. [PubMed]
16. Al-Manei, K.; Almotairy, N.; Bostanci, N.; Kumar, A.; Grigoriadis, A. Effect of Chewing on the Expression of Salivary Protein Composition: A Systematic Review. *Proteom. Clin. Appl.* **2020**, *14*, e1900039. [CrossRef] [PubMed]
17. Abrams, C.K.; Hamosh, M.; Hubbard, V.S.; Dutta, S.K.; Hamosh, P. Lingual lipase in cystic fibrosis. Quantitation of enzyme activity in the upper small intestine of patients with exocrine pancreatic insufficiency. *J. Clin. Investig.* **1984**, *73*, 374–382. [CrossRef]
18. Carriere, F.; Grandval, P.; Renou, C.; Palomba, A.; Prieri, F.; Giallo, J.; Henniges, F.; Sander-Struckmeier, S.; Laugier, R. Quantitative study of digestive enzyme secretion and gastrointestinal lipolysis in chronic pancreatitis. *Clin. Gastroenterol. Hepatol.* **2005**, *3*, 28–38. [CrossRef]
19. Stern, R.C.; Eisenberg, J.D.; Wagener, J.S.; Ahrens, R.; Rock, M.; doPico, G.; Orenstein, D.M. A comparison of the efficacy and tolerance of pancrelipase and placebo in the treatment of steatorrhea in cystic fibrosis patients with clinical exocrine pancreatic insufficiency. *Am. J. Gastroenterol.* **2000**, *95*, 1932–1938. [CrossRef]
20. Konstan, M.W.; Stern, R.C.; Trout, J.R.; Sherman, J.M.; Eigen, H.; Wagener, J.S.; Duggan, C.; Wohl, M.E.; Colin, P. Ultrase MT12 and Ultrase MT20 in the treatment of exocrine pancreatic insufficiency in cystic fibrosis: Safety and efficacy. *Aliment. Pharmacol. Ther.* **2004**, *20*, 1365–1371. [CrossRef] [PubMed]
21. Wooldridge, J.L.; Heubi, J.E.; Amaro-Galvez, R.; Boas, S.R.; Blake, K.V.; Nasr, S.Z.; Chatfield, B.; McColley, S.A.; Woo, M.S.; Hardy, K.A.; et al. EUR-1008 pancreatic enzyme replacement is safe and effective in patients with cystic fibrosis and pancreatic insufficiency. *J. Cyst. Fibros.* **2009**, *8*, 405–417. [CrossRef] [PubMed]
22. Trapnell, B.C.; Maguiness, K.; Graff, G.R.; Boyd, D.; Beckmann, K.; Caras, S. Efficacy and safety of Creon 24,000 in subjects with exocrine pancreatic insufficiency due to cystic fibrosis. *J. Cyst. Fibros.* **2009**, *8*, 370–377. [CrossRef]
23. Liao, T.H.; Hamosh, P.; Hamosh, M. Fat digestion by lingual lipase: Mechanism of lipolysis in the stomach and upper small intestine. *Pediatr. Res.* **1984**, *18*, 402–409. [CrossRef]

24. Kulkarni, B.V.; Mattes, R.D. Lingual lipase activity in the orosensory detection of fat by humans. *Am. J. Physiol. Regul. Integr. Comp. Physiol.* **2014**, *306*, R879–R885. [CrossRef]
25. Kulkarni, B.; Mattes, R. Evidence for presence of nonesterified fatty acids as potential gustatory signaling molecules in humans. *Chem. Senses* **2013**, *38*, 119–127. [CrossRef]
26. DeNigris, S.J.; Hamosh, M.; Kasbekar, D.K.; Lee, T.C.; Hamosh, P. Lingual and gastric lipases: Species differences in the origin of prepancreatic digestive lipases and in the localization of gastric lipase. *Biochim. Biophys. Acta* **1988**, *959*, 38–45. [CrossRef] [PubMed]
27. Abrams, C.K.; Hamosh, M.; Lee, T.C.; Ansher, A.F.; Collen, M.J.; Lewis, J.H.; Benjamin, S.B.; Hamosh, P. Gastric lipase: Localization in the human stomach. *Gastroenterology* **1988**, *95*, 1460–1464. [CrossRef]
28. Sams, L.; Paume, J.; Giallo, J.; Carriere, F. Relevant pH and lipase for in vitro models of gastric digestion. *Food Funct.* **2016**, *7*, 30–45. [CrossRef]
29. Carriere, F.; Renou, C.; Ransac, S.; Lopez, V.; De Caro, J.; Ferrato, F.; De Caro, A.; Fleury, A.; Sanwald-Ducray, P.; Lengsfeld, H.; et al. Inhibition of gastrointestinal lipolysis by Orlistat during digestion of test meals in healthy volunteers. *Am. J. Physiol. Gastrointest. Liver Physiol.* **2001**, *281*, G16–G28. [CrossRef] [PubMed]
30. Fernandez, S.; Jannin, V.; Rodier, J.D.; Ritter, N.; Mahler, B.; Carriere, F. Comparative study on digestive lipase activities on the self emulsifying excipient Labrasol, medium chain glycerides and PEG esters. *Biochim. Biophys. Acta* **2007**, *1771*, 633–640. [CrossRef]
31. Gargouri, Y.; Pieroni, G.; Lowe, P.A.; Sarda, L.; Verger, R. Human gastric lipase. The effect of amphiphiles. *Eur. J. Biochem.* **1986**, *156*, 305–310. [CrossRef] [PubMed]
32. Bernback, S.; Blackberg, L.; Hernell, O. The complete digestion of human milk triacylglycerol in vitro requires gastric lipase, pancreatic colipase-dependent lipase, and bile salt-stimulated lipase. *J. Clin. Investig.* **1990**, *85*, 1221–1226. [CrossRef] [PubMed]
33. Gargouri, Y.; Pieroni, G.; Riviere, C.; Lowe, P.A.; Sauniere, J.F.; Sarda, L.; Verger, R. Importance of human gastric lipase for intestinal lipolysis: An in vitro study. *Biochim. Biophys. Acta* **1986**, *879*, 419–423. [CrossRef]
34. MacGregor, I.L.; Martin, P.; Meyer, J.H. Gastric emptying of solid food in normal man and after subtotal gastrectomy and truncal vagotomy with pyloroplasty. *Gastroenterology* **1977**, *72*, 206–211. [CrossRef]
35. Mayer, E.A.; Thompson, J.B.; Jehn, D.; Reedy, T.; Elashoff, J.; Meyer, J.H. Gastric emptying and sieving of solid food and pancreatic and biliary secretion after solid meals in patients with truncal vagotomy and antrectomy. *Gastroenterology* **1982**, *83*, 184–192. [CrossRef]
36. Al Massadi, O.; Tschop, M.H.; Tong, J. Ghrelin acylation and metabolic control. *Peptides* **2011**, *32*, 2301–2308. [CrossRef]
37. Campos, L.A.; Sancho, J. The active site of pepsin is formed in the intermediate conformation dominant at mildly acidic pH. *FEBS Lett.* **2003**, *538*, 89–95. [CrossRef] [PubMed]
38. Hamuro, Y.; Coales, S.J.; Molnar, K.S.; Tuske, S.J.; Morrow, J.A. Specificity of immobilized porcine pepsin in H/D exchange compatible conditions. *Rapid Commun. Mass. Spectrom.* **2008**, *22*, 1041–1046. [CrossRef]
39. Conigrave, A.D.; Franks, A.H.; Brown, E.M.; Quinn, S.J. L-amino acid sensing by the calcium-sensing receptor: A general mechanism for coupling protein and calcium metabolism? *Eur. J. Clin. Nutr.* **2002**, *56*, 1072–1080. [CrossRef]
40. Taylor, I.L.; Byrne, W.J.; Christie, D.L.; Ament, M.E.; Walsh, J.H. Effect of individual l-amino acids on gastric acid secretion and serum gastrin and pancreatic polypeptide release in humans. *Gastroenterology* **1982**, *83*, 273–278. [CrossRef]
41. Hooton, D.; Lentle, R.; Monro, J.; Wickham, M.; Simpson, R. The Secretion and Action of Brush Border Enzymes in the Mammalian Small Intestine. *Rev. Physiol. Biochem. Pharmacol.* **2015**, *168*, 59–118. [CrossRef]
42. Date, K.; Satoh, A.; Iida, K.; Ogawa, H. Pancreatic alpha-Amylase Controls Glucose Assimilation by Duodenal Retrieval through N-Glycan-specific Binding, Endocytosis, and Degradation. *J. Biol. Chem.* **2015**, *290*, 17439–17450. [CrossRef]
43. Amiri, M.; Naim, H.Y. Characterization of Mucosal Disaccharidases from Human Intestine. *Nutrients* **2017**, *9*, 1106. [CrossRef]
44. Skovbjerg, H. Immunoelectrophoretic studies on human small intestinal brush border proteins--the longitudinal distribution of peptidases and disaccharidases. *Clin. Chim. Acta* **1981**, *112*, 205–212. [CrossRef] [PubMed]
45. Danialifar, T.F.; Chumpitazi, B.P.; Mehta, D.I.; Di Lorenzo, C. Genetic and acquired sucrase-isomaltase deficiency: A clinical review. *J. Pediatr. Gastroenterol. Nutr.* **2024**, *78*, 774–782. [CrossRef]
46. Zheng, T.; Eswaran, S.; Photenhauer, A.L.; Merchant, J.L.; Chey, W.D.; D'Amato, M. Reduced efficacy of low FODMAPs diet in patients with IBS-D carrying sucrase-isomaltase (SI) hypomorphic variants. *Gut* **2020**, *69*, 397–398. [CrossRef]
47. Gericke, B.; Amiri, M.; Scott, C.R.; Naim, H.Y. Molecular pathogenicity of novel sucrase-isomaltase mutations found in congenital sucrase-isomaltase deficiency patients. *Biochim. Biophys. Acta Mol. Basis Dis.* **2017**, *1863*, 817–826. [CrossRef] [PubMed]
48. Ellestad-Sayed, J.J.; Haworth, J.C.; Hildes, J.A. Disaccharide malabsorption and dietary patterns in two Canadian Eskimo communities. *Am. J. Clin. Nutr.* **1978**, *31*, 1473–1478. [CrossRef]
49. Uhrich, S.; Wu, Z.; Huang, J.Y.; Scott, C.R. Four mutations in the SI gene are responsible for the majority of clinical symptoms of CSID. *J. Pediatr. Gastroenterol. Nutr.* **2012**, *55* (Suppl. 2), S34–S35. [CrossRef] [PubMed]
50. Lee, B.H.; Hamaker, B.R. Maltase Has Most Versatile alpha-Hydrolytic Activity Among the Mucosal alpha-Glucosidases of the Small Intestine. *J. Pediatr. Gastroenterol. Nutr.* **2018**, *66* (Suppl. 3), S7–S10. [CrossRef]

51. El-Chammas, K.; Williams, S.E.; Miranda, A. Disaccharidase Deficiencies in Children with Chronic Abdominal Pain. *JPEN J. Parenter. Enter. Nutr.* **2017**, *41*, 463–469. [CrossRef]
52. Viswanathan, L.; Rao, S.S. Intestinal Disaccharidase Deficiency in Adults: Evaluation and Treatment. *Curr. Gastroenterol. Rep.* **2023**, *25*, 134–139. [CrossRef]
53. Oesterreicher, T.J.; Nanthakumar, N.N.; Winston, J.H.; Henning, S.J. Rat trehalase: cDNA cloning and mRNA expression in adult rat tissues and during intestinal ontogeny. *Am. J. Physiol.* **1998**, *274*, R1220–R1227. [CrossRef]
54. Gudmand-Hoyer, E.; Fenger, H.J.; Skovbjerg, H.; Kern-Hansen, P.; Madsen, P.R. Trehalase deficiency in Greenland. *Scand. J. Gastroenterol.* **1988**, *23*, 775–778. [CrossRef] [PubMed]
55. Murray, I.A.; Coupland, K.; Smith, J.A.; Ansell, I.D.; Long, R.G. Intestinal trehalase activity in a UK population: Establishing a normal range and the effect of disease. *Br. J. Nutr.* **2000**, *83*, 241–245. [CrossRef] [PubMed]
56. Savilahti, E.; Launiala, K.; Kuitunen, P. Congenital lactase deficiency. A clinical study on 16 patients. *Arch. Dis. Child.* **1983**, *58*, 246–252. [CrossRef]
57. Wanes, D.; Husein, D.M.; Naim, H.Y. Congenital Lactase Deficiency: Mutations, Functional and Biochemical Implications, and Future Perspectives. *Nutrients* **2019**, *11*, 461. [CrossRef] [PubMed]
58. Antonowicz, I.; Lebenthal, E. Developmental pattern of small intestinal enterokinase and disaccharidase activities in the human fetus. *Gastroenterology* **1977**, *72*, 1299–1303. [CrossRef]
59. Sterchi, E.E.; Mills, P.R.; Fransen, J.A.; Hauri, H.P.; Lentze, M.J.; Naim, H.Y.; Ginsel, L.; Bond, J. Biogenesis of intestinal lactase-phlorizin hydrolase in adults with lactose intolerance. Evidence for reduced biosynthesis and slowed-down maturation in enterocytes. *J. Clin. Investig.* **1990**, *86*, 1329–1337. [CrossRef]
60. Heyman, M.B.; Committee on Nutrition. Lactose intolerance in infants, children, and adolescents. *Pediatrics* **2006**, *118*, 1279–1286. [CrossRef]
61. Van Hoogevest, P.; Tiemessen, H.; Metselaar, J.M.; Drescher, S.; Fahr, A. The Use of Phospholipids to Make Pharmaceutical Form Line Extensions. *Eur. J. Lipid Sci. Technol.* **2021**, *124*, 2000297. [CrossRef]
62. Richmond, B.L.; Boileau, A.C.; Zheng, S.; Huggins, K.W.; Granholm, N.A.; Tso, P.; Hui, D.Y. Compensatory phospholipid digestion is required for cholesterol absorption in pancreatic phospholipase A(2)-deficient mice. *Gastroenterology* **2001**, *120*, 1193–1202. [CrossRef] [PubMed]
63. Nilsson, A.; Duan, R.D. Pancreatic and mucosal enzymes in choline phospholipid digestion. *Am. J. Physiol. Gastrointest. Liver Physiol.* **2019**, *316*, G425–G445. [CrossRef] [PubMed]
64. Rigtrup, K.M.; McEwen, L.R.; Said, H.M.; Ong, D.E. Retinyl ester hydrolytic activity associated with human intestinal brush border membranes. *Am. J. Clin. Nutr.* **1994**, *60*, 111–116. [CrossRef]
65. Duan, R.D.; Cheng, Y.; Hansen, G.; Hertervig, E.; Liu, J.J.; Syk, I.; Sjostrom, H.; Nilsson, A. Purification, localization, and expression of human intestinal alkaline sphingomyelinase. *J. Lipid Res.* **2003**, *44*, 1241–1250. [CrossRef]
66. Hertervig, E.; Nilsson, A.; Bjork, J.; Hultkrantz, R.; Duan, R.D. Familial adenomatous polyposis is associated with a marked decrease in alkaline sphingomyelinase activity: A key factor to the unrestrained cell proliferation? *Br. J. Cancer* **1999**, *81*, 232–236. [CrossRef]
67. Sjoqvist, U.; Hertervig, E.; Nilsson, A.; Duan, R.D.; Ost, A.; Tribukait, B.; Lofberg, R. Chronic colitis is associated with a reduction of mucosal alkaline sphingomyelinase activity. *Inflamm. Bowel Dis.* **2002**, *8*, 258–263. [CrossRef]
68. Lundgren, P.; Nilsson, A.; Duan, R.D. Distribution and properties of neutral ceramidase activity in rat intestinal tract. *Dig. Dis. Sci.* **2001**, *46*, 765–772. [CrossRef]
69. Curtis, K.J.; Gaines, H.D.; Kim, Y.S. Protein digestion and absorption in rats with pancreatic duct occlusion. *Gastroenterology* **1978**, *74*, 1271–1276. [CrossRef]
70. Mamone, G.; Picariello, G. Optimized extraction and large-scale proteomics of pig jejunum brush border membranes for use in in vitro digestion models. *Food Res. Int.* **2023**, *164*, 112326. [CrossRef]
71. Mina-Osorio, P. The moonlighting enzyme CD13: Old and new functions to target. *Trends Mol. Med.* **2008**, *14*, 361–371. [CrossRef]
72. Rohrborn, D.; Wronkowitz, N.; Eckel, J. DPP4 in Diabetes. *Front. Immunol.* **2015**, *6*, 386. [CrossRef]
73. Holmes, R.S.; Spradling-Reeves, K.D.; Cox, L.A. Mammalian Glutamyl Aminopeptidase Genes (ENPEP) and Proteins: Comparative Studies of a Major Contributor to Arterial Hypertension. *J. Data Min. Genom. Proteom.* **2017**, *8*, 2. [CrossRef]
74. Garcia-Guerrero, M.C.; Garcia-Pardo, J.; Berenguer, E.; Fernandez-Alvarez, R.; Barfi, G.B.; Lyons, P.J.; Aviles, F.X.; Huber, R.; Lorenzo, J.; Reverter, D. Crystal structure and mechanism of human carboxypeptidase O: Insights into its specific activity for acidic residues. *Proc. Natl. Acad. Sci. USA* **2018**, *115*, E3932–E3939. [CrossRef]
75. Stevens, B.R.; Phillips, M.I.; Fernandez, A. Ramipril inhibition of rabbit (*Oryctolagus cuniculus*) small intestinal brush border membrane angiotensin converting enzyme. *Comp. Biochem. Physiol. C Comp. Pharmacol. Toxicol.* **1988**, *91*, 493–497. [CrossRef]
76. Jaworska, K.; Koper, M.; Ufnal, M. Gut microbiota and renin-angiotensin system: A complex interplay at local and systemic levels. *Am. J. Physiol. Gastrointest. Liver Physiol.* **2021**, *321*, G355–G366. [CrossRef]

77. Penninger, J.M.; Grant, M.B.; Sung, J.J.Y. The Role of Angiotensin Converting Enzyme 2 in Modulating Gut Microbiota, Intestinal Inflammation, and Coronavirus Infection. *Gastroenterology* **2021**, *160*, 39–46. [CrossRef] [PubMed]
78. Holzinger, A.; Maier, E.M.; Buck, C.; Mayerhofer, P.U.; Kappler, M.; Haworth, J.C.; Moroz, S.P.; Hadorn, H.B.; Sadler, J.E.; Roscher, A.A. Mutations in the proenteropeptidase gene are the molecular cause of congenital enteropeptidase deficiency. *Am. J. Hum. Genet.* **2002**, *70*, 20–25. [CrossRef]
79. Green, N.; Chan, C.; Ooi, C.Y. The gastrointestinal microbiome, small bowel bacterial overgrowth, and microbiome modulators in cystic fibrosis. *Pediatr. Pulmonol.* **2024**, *59* (Suppl. 1), S70–S80. [CrossRef] [PubMed]
80. Caporilli, C.; Gianni, G.; Grassi, F.; Esposito, S. An Overview of Short-Bowel Syndrome in Pediatric Patients: Focus on Clinical Management and Prevention of Complications. *Nutrients* **2023**, *15*, 2341. [CrossRef] [PubMed]
81. Goulet, O.; Abi Nader, E.; Pigneur, B.; Lambe, C. Short Bowel Syndrome as the Leading Cause of Intestinal Failure in Early Life: Some Insights into the Management. *Pediatr. Gastroenterol. Hepatol. Nutr.* **2019**, *22*, 303–329. [CrossRef] [PubMed]
82. Drucker, D.J. Glucagon-like peptide 2. *J. Clin. Endocrinol. Metab.* **2001**, *86*, 1759–1764. [CrossRef] [PubMed]

biomolecules

MDPI

Review

Amylase Binding to Oral Streptococci: A Key Interaction for Human Oral Microbial Ecology, Adaptation and Fitness

Amarpreet Sabharwal [1], Elaine M. Haase [2] and Frank A. Scannapieco [2,*]

1 Schulich School of Medicine and Dentistry, Western University, London, ON N6A 5C1, Canada; amarpreet.sabharwal@schulich.uwo.ca
2 Department of Oral Biology, School of Dental Medicine, University at Buffalo, Buffalo, NY 14214, USA; emhaase6570@gmail.com
* Correspondence: fas1@buffalo.edu

Abstract: The interaction between human salivary alpha-amylase (HSAmy) and amylase-binding oral streptococci (ABS) helps determine the bacteria that colonize the oral cavity by establishing dental biofilms. Streptococci are important pioneer species of the oral cavity and influence oral health as well as common diseases such as dental caries. Various oral streptococcal species express distinct amylase-binding proteins, among which amylase-binding protein A (AbpA), encoded by the *abpA* gene in *Streptococcus gordonii* and several other species, which is the most extensively studied. Amylase binding facilitates microbial adhesion to host surfaces and biofilm formation and enables bacteria to harness the host's amylase enzymatic activity at their cell surface, enhancing their capacity to metabolize dietary starch for nutritional gain. Additionally, amylase binding may also influence bacterial cell division and stress tolerance by engaging novel bacterial signaling pathways. From an evolutionary perspective, both Neanderthals and modern humans exhibit functional adaptations in nutrient metabolism, including selection for salivary amylase-binding oral streptococci, highlighting the importance of microbial co-adaptation in response to host diet. Further research is warranted to elucidate the broader roles of amylase binding to bacteria in host-bacterial signaling, bacterial cell division and fitness and the evolutionary trajectory of the oral microbiome.

Keywords: dental plaque; saliva; biofilm; bacterial colonization; microbial evolution

1. Introduction

The human oral cavity represents a complex and dynamic ecosystem where the resident microbiome is constantly exposed to changing environmental conditions, including changes in pH, nutrients, salivary flow and composition and mechanical forces [1]. For effective colonization and persistence in the mouth, microbes have developed numerous strategies for adaptation to their environment. Commensal oral streptococci are pioneer colonizers of early dental biofilm on oral hard and soft tissues. Since dental biofilm formation is essential to the pathogenesis of dental caries and periodontal disease, understanding molecular mechanisms of host–microbe and microbe–microbe interactions may hope to inform future diagnostic and therapeutic approaches for these common disorders.

Saliva provides an important influence in oral microbial ecology [2]. One salivary component of particular significance in this regard is amylase, an abundant enzyme produced by serous cells of the parotid gland but also by sublingual, submaxillary and minor salivary glands. The amylase concentration in saliva varies between 0.04 and 0.4 mg/mL and may constitute up to 5% of the total salivary protein. Salivary stimulation can dramatically increase the levels of amylase. The enzyme's main catalytic specificity is to hydrolyze the alpha-1,4-glucosidic bonds of starch, glycogen and other polysaccharides, releasing maltose and maltodextrin that provides an abundant source of carbohydrate for oral bacterial nutrition.

Other interesting functions have been ascribed to amylase. An interaction between human salivary alpha-amylase (HSAmy) and certain oral streptococci has been well-documented and may play an important role in regulating the colonization of these bacteria in the mouth [3]. These amylase-binding streptococci (ABS) are abundant in the oral cavity. The interaction of HSAmy and ABS is mediated by amylase-binding protein(s) (ABP) [3]. Once bound to the bacterial surface, amylase retains enzymatic activity to mediate starch hydrolysis to fermentable oligosaccharides that contribute to bacterial nutrition [4,5]. The binding of amylase to the cell surface is inhibited by starch and maltotriose, but not maltose, supporting the involvement of specific receptor(s) on the bacterial surface. The addition of amylase to culture medium containing starch enhances the growth of *Streptococcus gordonii* [6].

A critical aspect of the host–microbe interaction in the mouth is the ecological success of early tooth colonizers (particularly oral streptococci), which must first interact with the salivary pellicle, a thin layer of saliva on oral structures. Amylase-binding protein A (AbpA) has been identified and localized to the surface of the bacteria to promote the adhesion of amylase-binding streptococci to amylase-coated surfaces in vitro, which serve as an analog of the salivary pellicle. AbpA-deficient mutants of these streptococci produce less biofilm than parental strains under in vitro flow conditions [6]. The binding of HSAmy to bacteria is calcium-independent, suggesting a mechanism distinct from enzymatic hydrolysis. The active site of HSAmy mediates saccharide binding and the hydrolysis of starch, with multiple secondary oligosaccharide-binding sites thought to enhance amylase affinity to starch granules [7]. These secondary oligosaccharide-binding sites have also been shown to play a role in bacterial binding. Mutation of aromatic residues in the secondary oligosaccharide-binding sites decreased the affinity of HSAmy binding to *S. gordonii*. Because amylase bound to bacteria retains enzymatic activity [7,8], the bacterium-binding site is likely distinct from that involved in enzymatic activity. Mutations in aromatic residues of secondary oligosaccharide-binding sites did not reduce the binding of amylase to hydroxyapatite, suggesting other un-identified sites mediating this interaction [7].

The possibility that amylase binding to oral streptococci might influence dental caries has been addressed by several studies [9–12]. Interestingly, Amylase-binding protein B (AbpB), rather than AbpA, appears to be more important for colonization of teeth in rats eating a starch-rich diet, and its deletion was partially masked if rats consumed a sucrose-starch diet. *S. gordonii* was compared to *Streptococcus mutans* with respect to oral colonization of the teeth and cariogenicity in a well-characterized rat model. Mutants of *S. gordonii* deficient in glucosyltransferase (GtfG), amylase-binding proteins (AbpA/AbpB), and *S. mutans* glucosyltransferase (GtfB) were studied. While both *S. gordonii* and *S. mutans* were abundant colonizers of rat's teeth in the presence of starch or sucrose diets, *S. mutans* always out-competed *S. gordonii* on the teeth. Caries induction reflects *S. mutans* or *S. gordonii* colonization abundance. *S. mutans* was found to be more cariogenic than *S. gordonii*. Thus, amylase binding to *S. gordonii* may not have a great influence on dental caries, especially when compared to the well-known cariogenic *S. mutans* exposed to a sucrose-rich diet.

The recent literature points towards a more complex role for the amylase-binding phenotype during the evolution of the human microbiome, where the emergence of ABS coincides with the introduction of plant starch in the human diet [13–16]. The goal of this review is to provide a summary of what is known to date of the relationship between HSAmy and ABS, and to provide suggestions for future studies.

2. The Genetic Basis and Diversity of ABPs

The ability of ABS to bind to HSAmy is not conferred by a single, conserved molecular mechanism. Instead, genetic and phylogenetic analyses suggest convergent evolution where several proteins, complemented by specialized processing systems, have adapted to perform a function that provides a selective advantage to the bacteria by interacting with an abundant host enzyme.

2.1. abpA-srtB Operon: System for Processing and Display in ABS

S. gordonii, a pioneer colonizer of tooth biofilm [17], expresses the protein AbpA [6], encoded by the *abpA* gene [18]. Genomic analysis has shown that it is part of a sophisticated expression and cell-wall anchoring system [19]. The *abpA* gene is co-transcribed with a proximally located downstream gene, *srtB*, that encodes a specialized class B sortase enzyme. The demonstration of co-transcription by PCR primers that span the gene junction confirms that *abpA-srtB* forms an operon for coordinated synthesis of the protein and the enzyme required for its cell wall localization [15]. Sortases are transpeptidases that covalently anchor surface proteins containing a conserved C-terminal sorting signal (e.g., LPXTG motif) to the peptidoglycan of Gram-positive bacteria [20]. While many bacteria possess a housekeeping sortase (like SrtA), which anchors a wide range of proteins, the AbpA-SrtB system in *S. gordonii* is distinct. The SrtB enzyme specifically recognizes a novel C-terminal cell-wall sorting motif within the AbpA protein [19]. The evolution of a dedicated sortase suggests a critical role for AbpA. A generic anchoring system may have been insufficient to ensure proper timing, density display, or conformation of AbpA on the cell surface, all of which are crucial for cell signaling roles. SrtB-mediated processing of AbpA can result in formation of a ladder profile in immunoblotting [19], suggesting that AbpA may undergo polymerization within the cell wall. The biological significance of this potential polymerization is not fully understood but seems specific for the processing of AbpA.

2.2. Multiple ABPs: Convergent Evolution and Horizontal Gene Transfer

Although the AbpA-SrtB system is well characterized in *S. gordonii*, it is not a universal mechanism for amylase binding among ABS. Extensive functional screening of diverse oral streptococcal species has shown a wide array of proteins that interact with amylase, varying in size from 20 to 87 kDa [21].

Phylogenetic and sequence analyses of ABPs have shown that they do not belong to a single, homologous protein family. ABPs cluster into at least six distinct and phylogenetically unrelated families: AbpA, AbpB and four novel families [21]. These novel protein families have been annotated based on homology to peptidoglycan-binding proteins, glutamine ABC transporters and choline-binding proteins. No single ancestral gene has been identified as an evolutionary source, suggesting convergent evolution by common environmental pressure from abundant salivary amylase in the oral cavity driving disparate genes towards a common functional solution. Further, comparative genomics of various streptococci provides evidence that acquisition of *abpA* was likely mediated by horizontal gene transfer [21].

3. Functional Characterization of ABPs

The functional significance of ABPs likely extends beyond roles in cell adhesion and nutrition. ABPs may also integrate environmental cues with cellular processes like cell division and stress tolerance by novel cell signaling pathways. Table 1 summarizes this evidence.

Table 1. Summary of evidence of amylase-binding proteins (ABPs) in oral ecology.

ABP families	
ABP family	Features
AbpA	~20 kDa; amylase-binding adhesin; SrtB anchored; novel sorting motif
AbpB	~87 kDa, dipeptidase; binds amylase but not essential for surface binding
Other	~20 to 87 kDa; homologous to peptidoglycan-binding proteins, glutamine ABC transporters and choline-binding proteins
AbpA-SrtB operon and processing	
Component	
abpA	Co-transcription with *srtB*
srtB	Dedicated sortase that anchors AbpA via novel motif; distinct from SrtA
Functional roles of ABPs	
Function	
Adhesion	AbpA enhances adhesion to HSAmy-coated surfaces
Nutrition	Starch is converted to maltose and maltotriose and AbpA⁻ mutants cannot grow on starch only
Regulation	Amylase triggers gene expression changes via AbpA; increase in fatty acid synthesis and stress tolerance
Cell division	Putative divisome links (FtsZ, PBP2b); septal localization; microscopic observation of AbpA at septum and chain length change in AbpA⁻ mutants
AbpA and oxidative stress	
Gene	
ccdA1	Redox homeostasis; downregulated in AbpA⁻ mutants
tlpA (etrx1)	Thioredoxin-like; downregulated in AbpA⁻ mutants
msrB (msrAB)	Repairs methionine sulfoxide; AbpA⁻ mutants sensitive to external H_2O_2
Histidine kinase/response regulator	Sense and regulate; operon-like; gradient downregulation
Ecological and evolutionary aspects	
Evidence	Observation and implication
AMY1 copy number	High-starch groups have increased AMY1 and salivary amylase; selects for ABS niche
Ancient human calculus	ABS present in Neanderthals/Late Pleistocene humans and predates agriculture
Non-human mammals	ABS abundant in several amylase-secretors; absent otherwise; physiologic adaptation of microbiome
S. suis analogs	SadP/ApuA (SrtA) with alpha-amylase activity; convergence without AbpA/SrtB

3.1. Canonical Functions: Adhesion and Nutrition

Following professional tooth cleaning, HSAmy adsorbs to hydroxyapatite to form part of the acquired enamel pellicle, creating a surface rich in receptors for ABS. AbpA has been shown in vitro to enhance the adhesion of *S. gordonii* to HSAmy-coated surfaces, a critical first step in early biofilm formation [6]. The AbpA-HSAmy interaction found to bind to amylase retains approximately 60% of its hydrolytic activity, allowing it to efficiently break down dietary starch into smaller, fermentable oligosaccharides like maltose and maltotriose directly at the cell surface [4,5,22]. The critical nature of this function is highlighted by the observation that preincubation of AbpA-deficient *S. gordonii* cells with salivary amylase, followed by washing in phosphate-buffered saline to remove unbound amylase, were unable to grow in a defined medium where starch was the sole carbohydrate source [6]. AbpB, annotated as a dipeptidase, also binds amylase, but it is not essential for amylase to bind to the cell surface of *S. gordonii* [23].

The roles of AbpA and AbpB in microbial ecology are complex (Table 1). While numerous in vitro studies demonstrate a role for AbpA in adhesion and biofilm formation, some in vivo studies show contradictory results. For example, a specific pathogen-free rat model of tooth colonization suggested that the expression of AbpA may, under certain conditions, inhibit colonization by *S. gordonii* [10]. Interestingly, when rats were inoculated with an AbpB mutant strain having intact AbpA, *S. gordonii* failed to colonize the teeth of starch-eating rats with abundant amylase in their saliva, but some colonization was restored with a starch/sucrose diet. Strains defective in AbpA colonized better than wild-type strains. AbpA appeared to inhibit colonization of the plaque biofilm in vivo. It was speculated that amylase/ABPs interact with glucosyltransferase or other colonization factors of these cells. These discrepancies suggest that the function of amylase binding by oral bacteria may be context dependent, with differences between initial adhesion to the host versus later stages of biofilm maturation and dispersal and further influenced by differences in host species.

3.2. Potential Roles for AbpA Beyond Adhesion and Nutrition

Microarray analyses examined the possibility that HSAmy binding to streptococci may affect genes involved in bacterial fitness [24]. Gene expression profiling showed that exposure to HSAmy elicits differential gene expression in wild-type *S. gordonii*, which is otherwise absent in an *abpA*⁻ mutant. Genes involved in fatty acid synthesis were upregulated and associated with increased bacterial growth, pH resistance and resistance to triclosan, all phenotypic changes enhancing bacterial survival in the oral environment.

To identify proteins that may interact with AbpA to affect differential gene expression, phage-display experiments were undertaken to screen for peptides that bind directly to AbpA (unpublished data, Table 2; see information in Supplementary Methods). The phage display was performed using a Ph.D-7 kit following the manufacturer's instructions. This unbiased approach showed potential interactions with some fatty acid-synthesis proteins previously identified, beta-ketoacyl-ACP reductase (FabG) [16], as well as with core components of the bacterial divisome [25], for example cell division protein FtsZ, an essential tubulin homolog that forms the foundational Z-ring supporting cell division [26] and penicillin-binding protein PBP2b, known to affect chain length in *S. pneumoniae* [27].

Table 2. Phage display sequences that potentially bind AbpA.

Sequence	Protein BLAST (2.7.1)	Max Score	Matched Sequence	SGO#	Recurrence
TSNNNLL	cell division protein (FtsZ)	21.8	SNNNLL	SGO_0675	(1/64 phages)
IVTQIPM	beta-ketoacyl-ACP reductase (FabG)	18	QIPM	SGO_1693	(1/64 phages)
TGSTRPW	beta-ketoacyl-ACP reductase (FabG)	17.6	TGSTR	SGO_1693	(4/64 phages)
EKKNMMN	sensor histidine kinase	16.8	+MMN	SGO_1174	(3/64 phages)
SSHSVQR	penicillin-binding protein 2B (PBP2b)	16.3	SHSVQ+R	SGO_1449	(20/64 phages)

That interactions between AbpA and the bacterial cell divisome proteins are possible is suggested by previous electron microscopic studies that showed localization of HSAmy to the cell division septum in dividing streptococci [28]. Additional studies were conducted of overnight cultures entering exponential phase of *S. gordonii* wild-type and mutant strains. After crystal violet staining, the bacteria were observed under a microscope with 1000 magnification. Bacterial chain lengths were determined by counting the cells in 10 random streptococcal chains per field (10 fields for each strain). The average cell number of 100 chains was calculated and subjected to statistical analysis. When AbpA is over expressed by a multicopy plasmid, not only is more amylase bound, but the average chain length of *S. gordonii* cells is significantly lengthened in both wild-type *S. gordonii* CH1 and in an isogenic *abpA*$^-$ mutant strain cultured in vitro (unpublished data, Figure 1a,b; see information in Supplementary Methods). This co-localization suggests that AbpA may be a transient member of the ABS divisome, potentially serving to link environmental sensing directly to bacterial cell metabolic regulation.

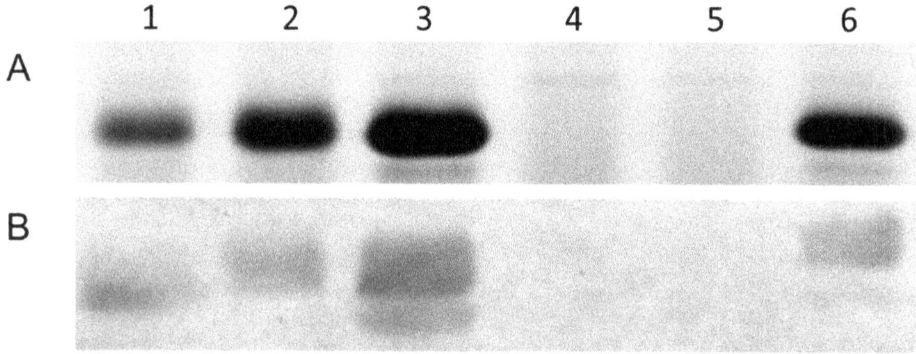

(**a**)

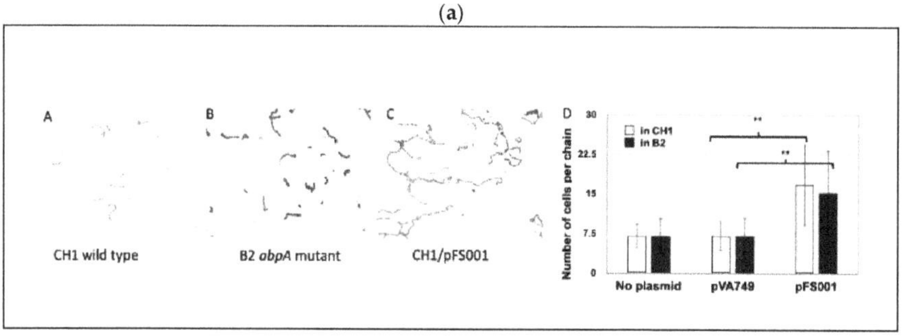

(**b**)

Figure 1. (**a**) Filter-concentrated supernatants from overnight TSBY cultures. Lane 1, CH1 (*S. gordonii* wild-type, parental strain); 2, CH1/pVA749 (parental strain with replicative plasmid); 3, CH1/pFS001 (parental strain with plasmid containing *abpA*); 4, B2 (AbpA-deficient CH1); 5, B2/pVA749 (AbpA-deficient CH1 with replicative plasmid); 6, B2/pFS001 (AbpA-deficient CH1 with plasmid containing *abpA*). A. Gel: Equal concentrations of protein were loaded onto 12% SDS-PAGE gel stained with AcquaStain (Bulldog Bio, Portsmouth, NH, USA). The major band of 20 kDa is AbpA. B. Far Western Blot: Proteins were transferred onto a PVDF membrane (Millipore-Sigma, Burlington, MA, USA), blocked with 3% skim milk in Tris-buffered saline with 0.1% Tween 20, incubated with purified human non-glycosylated salivary alpha-amylase, followed by primary antibody (rabbit anti-human amylase) and secondary antibody (goat anti-rabbit IgG conjugated with alkaline phosphatase (BioRad). Color was developed with nitro blue tetrazolium (NBT; Millipore-Sigma, Burlington, MA, USA). (**b**) Chain length of cells in mid-log phase. A, B, C, Gram stain. D, number of cells per chain. Strains: CH1 (wild type), B2 (*abpA*$^-$ mutant), B2/pFS001 (complemented mutant). Vectors: pVA749 (empty vector); pFS001 (pVA749 with *abpA*). ** Statistically significant ($p < 0.05$).

3.3. AbpA-CTM Axis: Pathway for Oxidative Stress Resistance

Comparative transcriptomic analysis of wild-type *S. gordonii* and an isogenic *abpA*⁻ mutant showed that a specific gene cluster (*ccdA1/tlpA/msrB*), homologous to the CTM gene cluster (*ccdA1/etrx1/msrAB2*) in *S. pneumoniae*, was consistently and significantly downregulated in the *abpA*⁻ mutant across all growth conditions, even in the absence of HSAmy [15]. This constitutive, HSAmy-independent effect demonstrates that AbpA may play an important role in regulating this gene cluster, beyond the response to its ligand.

The CTM gene cluster in *S. pneumoniae* is involved in redox homeostasis and stress response [29,30]. The core components of this CTM homologous system in *S. gordonii* encode a cytochrome c-type biogenesis protein (CcdA1, a thioredoxin-like protein TlpA/etrx1 homolog) and a peptide methionine sulfoxide reductase (MsrAB homolog). The function of these components indicates a role in the management of oxidative stress. Methionine sulfoxide reductases (MsrA and MsrB) are important for ABSs that lack catalase [31]. These enzymes repair oxidatively damaged proteins by reducing methionine sulfoxide to methionine [32]. The functional outcome of this oxidative stress protection was tested on an *abpA*⁻ mutant of *S. gordonii*, which was found to be significantly more sensitive to killing by extracellularly applied hydrogen peroxide when compared to an *abpA*-complemented strain [15]. Importantly, the mutant did not show an increased sensitivity to oxidative stress generated internally (by paraquat treatment) or to the oxidative burst within phagocytic cells (unpublished data; see information in Supplementary Methods). This distinction is important as it indicates that the AbpA-regulated CTM system may have evolved to defend against external oxidative injuries such as hydrogen peroxide produced by competing bacteria in oral biofilm.

The CTM gene cluster in *S. gordonii* and *S. pneumoniae* is flanked at the C-terminus sequence coding for a sensor histidine kinase and a response regulator. The histidine kinase and response regulator are homologous to the two-component system (TCS) in *S. pneumoniae* that is thought to be associated with a response to environmental stress [33] and multiple TCSs have been shown to regulate responses to environmental stresses in *S. gordonii* [34]. PCR studies in *S. gordonii* using intergenic primers suggest that the CTM, sensor histidine kinase and cognate response regulator genes may be transcribed as a single, long polycistronic message. The expression of this putative operon is influenced by AbpA as all genes in this cluster show some level of downregulation in the *abpA*⁻ mutant. Interestingly, this downregulation occurs in a gradient, with the most upstream gene (*ccdA1*) showing the greatest decrease in expression and the most downstream gene (sensor histidine kinase) showing the least. This suggests a complex regulatory architecture possibly involving transcriptional polarity or attenuation within the operon. Phage display experiments raised the possibility of a potential physical interaction between AbpA and the sensor histidine kinase in this TCS. A direct signaling pathway in which AbpA (after conformational change or a cell surface interaction) directly modulates the activity of the sensor histidine kinase which then phosphorylates the response regulator to control transcription of the entire CTM-TCS locus is possible.

The role of an AbpA-CTM axis for defense against extracellular oxidative stress should be qualified against its ecological purpose. Pioneer colonizers like *S. gordonii* produce hydrogen peroxide via the *spxB*-encoded pyruvate oxidase to inhibit growth of other species in competition such as the cariogenic pathogen, *S. mutans* [35]. Constitutive upregulation by AbpA may allow ABS to withstand the oxidative conditions they generate to establish and defend themselves in the oral biofilm. In this regard, CTM may not be viewed as a generic stress-response system but as a defensive component to complement an active strategy by pioneer colonizers to establish superiority early in an ecological system.

4. Ecological and Evolutionary Considerations

Molecular mechanisms governing the binding of HSAmy to ABS are the result of a long and dynamic co-evolutionary history with their mammalian hosts, shaped by dietary shifts, ecological competition and constant pressure to adapt and survive in a complex oral biofilm.

4.1. Co-Evolution: Interactions of Diet, Genes and Microbiome

This host–microbe co-adaptation is not a recent event. Analysis of dental calculus from ancient hominids, including Neanderthals and Late Pleistocene modern humans, revealed the presence of substantial proportions of amylase-binding oral streptococci [13,14]. This indicates that the selective pressures for HSAmy-ABS like interactions predate agriculture by tens of thousands of years, likely linked to earlier dietary shifts that incorporated starch-rich foods such as tubers into the hominid diet. A prominent event in human evolution was a significant increase in starch consumption, particularly following the Neolithic revolution and the advent of agriculture [36]. This dietary shift likely created a strong selective pressure for the extraction of energy from starch-rich foods and is reflected by the expansion of AMY1 gene copy numbers [37]. Populations with a history of high-starch diets tend to have more AMY1 copies and consequently higher concentrations of amylase in their saliva [36,38]. This change in host physiology likely had a persistent effect on the ecology of the oral microbiome [39]. The increased availability of HSAmy and its starch-derived byproducts created a new and nutrient-rich niche. Oral streptococci that evolved and acquired the ability to bind to HSAmy enjoyed a substantial fitness advantage.

4.2. Amylase-Binding Bacteria in Other Mammals

The HSAmy-to-ABS binding phenotype is not restricted to the human oral microbiome. A comparative study of dental biofilms in 14 non-human mammals showed ABS ranging from 2 to 31% of total microbiota in five out of six amylase-secreting animals [40]. Pigs, which secrete amylase, were a notable exception. All animals that did not secrete amylase in their saliva did not have ABS. *Streptococcus suis* (*S. suis*) is an important colonizer of pig tonsils and must adhere to host cells for persistence and infection [41]. Adhesion proteins streptococcal adhesin P (SadP) and amylopullulanase A (ApuA) are both processed by SrtA for cell wall anchoring [42], and ApuA has alpha-amylase activity similar to AbpA [43]. A SrtA knockout can reduce adhesion in S. suis [44]. While AbpA is processed and anchored to the cell wall by SrtB in ABS and there is lack of evidence of a homologous AbpA/SrtB system in *S. suis*, it is likely that the carbohydrate-based adhesion in *S. suis* is the predominant mechanism of adhesion [42]. A combination of gene functions in distinct but related bacteria can account for oral microbiome adaptation in diverse hosts with similar ecological demands, i.e., oral cavities in humans and animals.

5. Future Directions

The major amylase-binding protein, AbpA, is co-transcribed with a proximally encoded novel sortase B. The potential multifunctionality of AbpA suggests it to be a novel protein by which bacteria may sense carbon sources and oxidative stress and react to a rapidly changing environment. In addition, the interaction of amylase with oral streptococci offers a unique model for host–protein interactions with commensal microflora. Fluctuations in amylase concentration in saliva could modulate the numbers of these organisms in plaque, in part through the mechanisms described above, with consequent effects on oral disease susceptibility. Future studies should further explore the HSAmy-AbpA interaction that may impact bacterial fitness to devise innovative approaches to control oral biofilms. The role of AbpA as a potential member of the streptococcal divisome also deserves attention. Additional studies are required to understand the role of amylase, ABS and starch in oral diseases such as dental caries and periodontal disease. Finally, the finding that the emergence of ABS as prominent members of the oral microbiome coincides with the adoption of starch in the human diet emphasizes the importance of this interaction in the evolutionary context and deserves further study.

It may be possible to design analogs that serve as inhibitors or promoters of microbial colonization and/or pathogenicity, with possible impact on disease prevention or control. This knowledge could extend beyond oral biofilm formation with potential application to other microbial communities affecting systemic health and disease.

In summary, the binding of HSAmy to AbpA appears to have multiple implications for the adhesion of bacteria to the host, the metabolism of dietary starch for bacterial nutrition and impacts cell division and signaling (Figure 2).

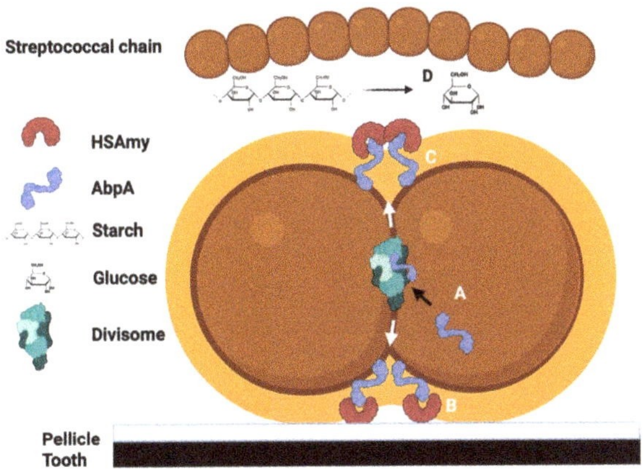

Figure 2. A summary of the potential roles for the HSAmy-AbpA interaction. A. AbpA is expressed, incorporated into the cell divisome and then released to the cell surface at the cell division septum. B. AbpA binds to HSAmy within salivary pellicle to promote bacterial adhesion to the tooth. C. HSAmy binds soluble enzymatically active HSAmy. D. Dietary starch is hydrolyzed, releasing simple sugars that can be utilized by the bacteria for energy.

6. Conclusions

The interaction of amylase on the surface of common and abundant oral streptococcal species likely confers an important adaptation impacting the fitness of these organisms within the host. A sophisticated system appears to have evolved to bind amylase to the surface of oral streptococci and possibly mediate signals from the host to the bacteria that influence cell division, growth, adhesion and colonization. The bacterial amylase-binding phenotype has been an important ecological determinant during the evolution of the oral microbiome. Further studies are required to determine the potential role of amylase binding in bacterial cell-signaling, cell division, overall bacterial fitness, oral disease pathogenesis and the evolution of the oral microbiome.

Supplementary Materials: The following supporting information can be downloaded at: https://www.mdpi.com/article/10.3390/biom15111616/s1, Methodology; original blot. Refs. [21,45–51] are cited in Supplementary Materials.

Author Contributions: Conceptualization, A.S. and F.A.S.; writing—original draft preparation, A.S. and F.A.S. writing—review and editing, A.S., E.M.H. and F.A.S. All authors have read and agreed to the published version of the manuscript.

Funding: Much of the research described by the authors' laboratory was funded in part by USPHS-NIH 1R01DE022673-01 from the National Institutes of Health, USA.

Institutional Review Board Statement: Not applicable.

Informed Consent Statement: Informed consent was obtained from all subjects involved in the studies described.

Data Availability Statement: The original contributions presented in this study are included in the article/Supplementary Material. Further inquiries can be directed to the corresponding authors.

Acknowledgments: The authors thank all of the students, post-doctoral fellows and other colleagues who contributed to these studies. Many thanks to Jiachuan Pan, M. Margaret Vickerman, Jason Kay for their contributions to the methods supplement.

Conflicts of Interest: The authors declare no conflicts of interest.

Abbreviations

The following abbreviations are used in this manuscript:

ABS	Amylase-Binding Streptococci
HSAmy	Human Alpha-amylase
AbpA	Amylase-binding Protein A
AbpB	Amylase-binding Protein B

References

1. Scannapieco, F.A.; Ruhl, S. The Oral Environment. In *Oral Microbiology and Immunology*, 3rd ed.; Richard, J., Lamont, G.N.H., Koo, H., Jenkinson, H.F., Eds.; ASM Press: Washington, DC, USA, 2019.
2. Scannapieco, F.A. Saliva-bacterium interactions in oral microbial ecology. *Crit. Rev. Oral Biol. Med.* **1994**, *5*, 203–248. [CrossRef] [PubMed]
3. Nikitkova, A.E.; Haase, E.M.; Scannapieco, F.A. Taking the starch out of oral biofilm formation: Molecular basis and functional significance of salivary alpha-amylase binding to oral streptococci. *Appl. Environ. Microbiol.* **2013**, *79*, 416–423. [CrossRef] [PubMed]
4. Douglas, C.W. The binding of human salivary alpha-amylase by oral strains of streptococcal bacteria. *Arch. Oral Biol.* **1983**, *28*, 567–573. [CrossRef]
5. Scannapieco, F.A.; Bergey, E.J.; Reddy, M.S.; Levine, M.J. Characterization of salivary alpha-amylase binding to *Streptococcus sanguis*. *Infect. Immun.* **1989**, *57*, 2853–2863. [CrossRef]
6. Rogers, J.D.; Palmer, R.J., Jr.; Kolenbrander, P.E.; Scannapieco, F.A. Role of *Streptococcus gordonii* amylase-binding protein A in adhesion to hydroxyapatite, starch metabolism, and biofilm formation. *Infect. Immun.* **2001**, *69*, 7046–7056. [CrossRef] [PubMed]
7. Ragunath, C.; Manuel, S.G.; Venkataraman, V.; Sait, H.B.; Kasinathan, C.; Ramasubbu, N. Probing the role of aromatic residues at the secondary saccharide-binding sites of human salivary alpha-amylase in substrate hydrolysis and bacterial binding. *J. Mol. Biol.* **2008**, *384*, 1232–1248. [CrossRef]
8. Scannapieco, F.A.; Bhandary, K.; Ramasubbu, N.; Levine, M.J. Structural relationship between the enzymatic and streptococcal binding sites of human salivary alpha-amylase. *Biochem. Biophys. Res. Commun.* **1990**, *173*, 1109–1115. [CrossRef]
9. Tanzer, J.M.; Baranowski, L.K.; Rogers, J.D.; Haase, E.M.; Scannapieco, F.A. Oral colonization and cariogenicity of *Streptococcus gordonii* in specific pathogen-free TAN:SPFOM(OM)BR rats consuming starch or sucrose diets. *Arch. Oral Biol.* **2001**, *46*, 323–333. [CrossRef]
10. Tanzer, J.M.; Grant, L.; Thompson, A.; Li, L.; Rogers, J.D.; Haase, E.M.; Scannapieco, F.A. Amylase-binding proteins A (AbpA) and B (AbpB) differentially affect colonization of rats' teeth by *Streptococcus gordonii*. *Microbiology* **2003**, *149 Pt 9*, 2653–2660. [CrossRef]
11. Tanzer, J.M.; Thompson, A.M.; Grant, L.P.; Vickerman, M.M.; Scannapieco, F.A. *Streptococcus gordonii's* sequenced strain CH1 glucosyltransferase determines persistent but not initial colonization of teeth of rats. *Arch. Oral Biol.* **2008**, *53*, 133–140. [CrossRef]
12. Tanzer, J.M.; Thompson, A.; Sharma, K.; Vickerman, M.M.; Haase, E.M.; Scannapieco, F.A. *Streptococcus mutans* out-competes *Streptococcus gordonii* in vivo. *J. Dent. Res.* **2012**, *91*, 513–519. [CrossRef]
13. Fellows Yates, J.A.; Velsko, I.M.; Aron, F.; Posth, C.; Hofman, C.A.; Austin, R.M.; Parker, C.E.; Mann, A.E.; Nagele, K.; Arthur, K.W.; et al. The evolution and changing ecology of the African hominid oral microbiome. *Proc. Natl. Acad. Sci. USA* **2021**, *118*, e2021655118. [CrossRef] [PubMed]

14. Velsko, I.M.; Warinner, C. Streptococcus abundance and oral site tropism in humans and non-human primates reflects host and lifestyle differences. *NPJ Biofilms Microbiomes* **2025**, *11*, 19. [CrossRef] [PubMed]
15. Haase, E.M.; Feng, X.; Pan, J.; Miecznikowski, J.C.; Scannapieco, F.A. Dynamics of the *Streptococcus gordonii* Transcriptome in Response to Medium, Salivary alpha-Amylase, and Starch. *Appl. Environ. Microbiol.* **2015**, *81*, 5363–5374. [CrossRef]
16. Nikitkova, A.E.; Haase, E.M.; Vickerman, M.M.; Gill, S.R.; Scannapieco, F.A. Response of fatty acid synthesis genes to the binding of human salivary amylase by *Streptococcus gordonii*. *Appl. Environ. Microbiol.* **2012**, *78*, 1865–1875. [CrossRef]
17. Kreth, J.; Merritt, J.; Qi, F. Bacterial and host interactions of oral streptococci. *DNA Cell Biol.* **2009**, *28*, 397–403. [CrossRef]
18. Rogers, J.D.; Haase, E.M.; Brown, A.E.; Douglas, C.W.; Gwynn, J.P.; Scannapieco, F.A. Identification and analysis of a gene (abpA) encoding a major amylase-binding protein in *Streptococcus gordonii*. *Microbiology* **1998**, *144 Pt 5*, 1223–1233. [CrossRef]
19. Liang, X.; Liu, B.; Zhu, F.; Scannapieco, F.A.; Haase, E.M.; Matthews, S.; Wu, H. A distinct sortase SrtB anchors and processes a streptococcal adhesin AbpA with a novel structural property. *Sci. Rep.* **2016**, *6*, 30966. [CrossRef]
20. Ton-That, H.; Marraffini, L.A.; Schneewind, O. Protein sorting to the cell wall envelope of Gram-positive bacteria. *Biochim. Biophys. Acta* **2004**, *1694*, 269–278. [CrossRef] [PubMed]
21. Haase, E.M.; Kou, Y.; Sabharwal, A.; Liao, Y.C.; Lan, T.; Lindqvist, C.; Scannapieco, F.A. Comparative genomics and evolution of the amylase-binding proteins of oral streptococci. *BMC Microbiol.* **2017**, *17*, 94.
22. Douglas, C.W.; Pease, A.A.; Whiley, R.A. Amylase-binding as a discriminator among oral streptococci. *FEMS Microbiol. Lett.* **1990**, *54*, 193–197. [CrossRef] [PubMed]
23. Li, L.; Tanzer, J.M.; Scannapieco, F.A. Identification and analysis of the amylase-binding protein B (AbpB) and gene (abpB) from *Streptococcus gordonii*. *FEMS Microbiol. Lett.* **2002**, *212*, 151–157. [CrossRef] [PubMed]
24. Nikitkova, A.E.; Haase, E.M.; Scannapieco, F.A. Effect of starch and amylase on the expression of amylase-binding protein A in *Streptococcus gordonii*. *Mol. Oral Microbiol.* **2012**, *27*, 284–294. [CrossRef]
25. Briggs, N.S.; Bruce, K.E.; Naskar, S.; Winkler, M.E.; Roper, D.I. The pneumococcal divisome: Dynamic control of *Streptococcus pneumoniae* cell division. *Front. Microbiol.* **2021**, *12*, 737396. [CrossRef] [PubMed]
26. Fleurie, A.; Lesterlin, C.; Manuse, S.; Zhao, C.; Cluzel, C.; Lavergne, J.P.; Franz-Wachtel, M.; Macek, B.; Combet, C.; Kuru, E.; et al. MapZ marks the division sites and positions FtsZ rings in *Streptococcus pneumoniae*. *Nature* **2014**, *516*, 259–262. [CrossRef]
27. Berg, K.H.; Stamsas, G.A.; Straume, D.; Havarstein, L.S. Effects of low PBP2b levels on cell morphology and peptidoglycan composition in *Streptococcus pneumoniae* R6. *J. Bacteriol.* **2013**, *195*, 4342–4354. [CrossRef]
28. Scannapieco, F.A.; Haraszthy, G.G.; Cho, M.I.; Levine, M.J. Characterization of an amylase-binding component of *Streptococcus gordonii* G9B. *Infect. Immun.* **1992**, *60*, 4726–4733. [CrossRef]
29. Andisi, V.F.; Hinojosa, C.A.; de Jong, A.; Kuipers, O.P.; Orihuela, C.J.; Bijlsma, J.J. Pneumococcal gene complex involved in resistance to extracellular oxidative stress. *Infect. Immun.* **2012**, *80*, 1037–1049. [CrossRef]
30. Saleh, M.; Bartual, S.G.; Abdullah, M.R.; Jensch, I.; Asmat, T.M.; Petruschka, L.; Pribyl, T.; Gellert, M.; Lillig, C.H.; Antelmann, H.; et al. Molecular architecture of *Streptococcus pneumoniae* surface thioredoxin-fold lipoproteins crucial for extracellular oxidative stress resistance and maintenance of virulence. *EMBO Mol. Med.* **2013**, *5*, 1852–1870. [CrossRef]
31. Jalal, N.; Lee, S.F. The MsrAB reducing pathway of *Streptococcus gordonii* is needed for oxidative stress tolerance, biofilm formation, and oral colonization in mice. *PLoS ONE* **2020**, *15*, e0229375. [CrossRef]
32. Kim, H.Y. The methionine sulfoxide reduction system: Selenium utilization and methionine sulfoxide reductase enzymes and their functions. *Antioxid. Redox Signal* **2013**, *19*, 958–969. [CrossRef] [PubMed]
33. Landete, J.M. Effector Molecules and Regulatory Proteins: Applications. *Trends Biotechnol.* **2016**, *34*, 777–780. [CrossRef]
34. Liu, Y.; Burne, R.A. Multiple two-component systems modulate alkali generation in *Streptococcus gordonii* in response to environmental stresses. *J. Bacteriol.* **2009**, *191*, 7353–7362. [CrossRef]
35. Zhu, L.; Kreth, J. The role of hydrogen peroxide in environmental adaptation of oral microbial communities. *Oxid. Med. Cell Longev.* **2012**, *2012*, 717843. [CrossRef]
36. Bolognini, D.; Halgren, A.; Lou, R.N.; Raveane, A.; Rocha, J.L.; Guarracino, A.; Soranzo, N.; Chin, C.S.; Garrison, E.; Sudmant, P.H. Recurrent evolution and selection shape structural diversity at the amylase locus. *Nature* **2024**, *634*, 617–625. [CrossRef]
37. Perry, G.H.; Dominy, N.J.; Claw, K.G.; Lee, A.S.; Fiegler, H.; Redon, R.; Werner, J.; Villanea, F.A.; Mountain, J.L.; Misra, R.; et al. Diet and the evolution of human amylase gene copy number variation. *Nat. Genet.* **2007**, *39*, 1256–1260. [CrossRef]
38. Pajic, P.; Pavlidis, P.; Dean, K.; Neznanova, L.; Romano, R.A.; Garneau, D.; Daugherity, E.; Globig, A.; Ruhl, S.; Gokcumen, O. Independent amylase gene copy number bursts correlate with dietary preferences in mammals. *eLife* **2019**, *8*, e44628. [CrossRef]
39. Adler, C.J.; Dobney, K.; Weyrich, L.S.; Kaidonis, J.; Walker, A.W.; Haak, W.; Bradshaw, C.J.; Townsend, G.; Soltysiak, A.; Alt, K.W.; et al. Sequencing ancient calcified dental plaque shows changes in oral microbiota with dietary shifts of the Neolithic and Industrial revolutions. *Nat. Genet.* **2013**, *45*, 450–455.e1. [CrossRef]
40. Scannapieco, F.A.; Solomon, L.; Wadenya, R.O. Emergence in human dental plaque and host distribution of amylase-binding streptococci. *J. Dent. Res.* **1994**, *73*, 1627–1635. [CrossRef] [PubMed]

41. Votsch, D.; Willenborg, M.; Weldearegay, Y.B.; Valentin-Weigand, P. *Streptococcus suis*—The "Two Faces" of a Pathobiont in the Porcine Respiratory Tract. *Front. Microbiol.* **2018**, *9*, 480. [CrossRef] [PubMed]
42. Kouki, A.; Haataja, S.; Loimaranta, V.; Pulliainen, A.T.; Nilsson, U.J.; Finne, J. Identification of a novel streptococcal adhesin P (SadP) protein recognizing galactosyl-alpha1-4-galactose-containing glycoconjugates: Convergent evolution of bacterial pathogens to binding of the same host receptor. *J. Biol. Chem.* **2011**, *286*, 38854–38864. [CrossRef]
43. Ferrando, M.L.; Fuentes, S.; de Greeff, A.; Smith, H.; Wells, J.M. ApuA, a multifunctional alpha-glucan-degrading enzyme of *Streptococcus suis*, mediates adhesion to porcine epithelium and mucus. *Microbiology* **2010**, *156 Pt 9*, 2818–2828. [CrossRef]
44. Wang, C.; Li, M.; Feng, Y.; Zheng, F.; Dong, Y.; Pan, X.; Cheng, G.; Dong, R.; Hu, D.; Feng, X.; et al. The involvement of sortase A in high virulence of STSS-causing *Streptococcus suis* serotype 2. *Arch. Microbiol.* **2009**, *191*, 23–33. [CrossRef]
45. Jenkinson, H.F. Cell-surface proteins of Streptococcus sanguis associated with cell hydrophobicity and coaggregation properties. *J. Gen. Microbiol.* **1986**, *132*, 1575–1589. [CrossRef]
46. Loo, C.Y.; Corliss, D.A.; Ganeshkumar, N. *Streptococcus gordonii* biofilm formation: Identification of genes that code for biofilm phenotypes. *J. Bacteriol.* **2000**, *182*, 1374–1382. [CrossRef] [PubMed]
47. Rogers, J.D.; Scannapieco, F.A. Catabolite repression and regulation of the amylase-binding protein gene of Streptococcus gordonii. *J. Bacteriol.* **2001**, *183*, 3521–3525. [CrossRef]
48. Sulavik, M.C.; Tardif, G.; Clewell, D.B. Identification of a gene, rgg, which regulates expression of glucosyltransferase and influences the Spp phenotype of Streptococcus gordonii Challis. *J. Bacteriol.* **1992**, *174*, 3577–3586. [CrossRef] [PubMed]
49. Jakubovics, N.S.; Smith, A.W.; Jenkinson, H.F. Oxidative stress tolerance is manganese (Mn(2+)) regulated in Streptococcus gordonii. *Microbiology* **2002**, *148 Pt 10*, 3255–3263. [CrossRef]
50. Flannagan, R.S.; Jaumouille, V.; Grinstein, S. The cell biology of phagocytosis. *Annu. Rev. Pathol.* **2012**, *7*, 61–98. [CrossRef]
51. Vaudaux, P.; Waldvogel, F.A. Gentamicin antibacterial activity in the presence of human polymorphonuclear leukocytes. *Antimicrob. Agents Chemother.* **1979**, *16*, 743–749. [CrossRef] [PubMed]

biomolecules

MDPI

Article

Hemodynamic and Morpho-Biochemical Parameters of Rabbit Blood After Injection of Enzyme Preparations

V. G. Vertiprakhov *, N. A. Sergeenkova , S. V. Karamushkina and B. Sh. Dashieva *

Department of Physiology, Ethology and Biochemistry of Animals, Russian State Agrarian University—Moscow Timiryazev Agricultural Academy, 49, Timiryazevskaya St., Moscow 127550, Russia; nsergeenkova@rgau-msha.ru (N.A.S.); sveta.vetmed@mail.ru (S.V.K.)

* Correspondence: vertiprakhov63@mail.ru (V.G.V.); dashieva.b.sh@rgau-msha.ru (B.S.D.)

Abstract: The anti-inflammatory effect of trypsin in animals and humans is the basis for the development of new veterinary and medical drugs and alternatives to antibiotics. The current experiment analyzed the effect of pig pancreatic tissue lyophilizate and crystalline trypsin on the hemodynamic and morpho-biochemical parameters of rabbit blood. The experiments were carried out on 20 rabbits of the Soviet chinchilla breed of 6–8 months of age. Animals were intramuscularly injected with sterile solution of 0.9% NaCl in 0.5 mL (group 1, n = 5), sterile solution of crystalline trypsin in 0.9% NaCl at a concentration of 0.25 mg/kg body weight (group 2, n = 5), sterile solution of crystalline trypsin in 0, 9% NaCl at a concentration of 0.5 mg/kg body weight (group 3, n = 5), or sterile suspension of pig pancreas lyophilizate at a concentration of 1 mg/kg body weight (group 4, n = 5). Animals were injected once daily for five consecutive days. Significant changes in arterial blood pressure, serum enzymes activity, and the count of various blood cellular components were induced by the administration of different trypsin preparations. All data obtained indicate the presence of a biologically active substance in the lyophilizate, the effect of which requires further animal studies to create a prototype for the development of new drugs for human and animal use.

Keywords: trypsin crystalline; lyophilizate pancreas pigs; rabbits; blood pressure; blood trypsin

1. Introduction

The problem with finding efficient alternatives to antibiotics is extremely relevant today. Trypsin is known to have anti-inflammatory and decongestant effects [1–3], which in its effectiveness can be compared to nonsteroidal, anti-inflammatory drugs. The use of enzyme preparations from animal tissue, including those prepared from the pancreas, has limited use in animal husbandry practice, since the crystalline trypsin is mainly used to treat wounds and inflammatory processes. The relevance of studying pancreatic proteases, secreted by digestive glands into the blood, is important due to the large number of studies performed in recent years on receptors activated by proteases. Receptors are located on the cell membranes of various organs and tissues, through which the functional activity of these organs and tissues can be enhanced or reduced under the influence of pancreatic proteases [4,5]. Trypsin has been found to be one of the activators of PAR receptors [6]. A study by Koshikawa N., Hasegawa S., Nagashima Y. et al. (1998) [7] showed trypsin expression in human and mouse non-pancreatic tissues. The trypsin gene was found to be expressed at high levels in the pancreas, spleen, and small intestine. In situ hybridization and immunohistochemistry have shown that trypsin is widely expressed in the epithelial cells of the skin, esophagus, stomach, small intestine, lungs, kidneys, liver, and extrahepatic bile ducts, as well as in the spleen and neurons [7]. Thus, the enzyme trypsin is often considered to be a hormone-like substance [8], which, when entering the blood, can have a regulatory effect on the behavior of animals [9], changes in blood pressure and heart rate in rabbits [10], as well as on metabolism in general [11].

The production of enzyme preparations for parenteral use depends not only on the technology, but also on the raw materials from which the enzyme is isolated. The pharmacopeial preparation, trypsin crystalline, is obtained from the pancreas of cattle. There is evidence that the proteolytic activity in the homogenate of the pancreas of cattle is 405 ± 4.9 mg casein/g pancreas tissue/min [10]. Protease activity reaches 839 ± 87.4 mg/g/min in the pancreatic tissue of broiler chickens and 317 ± 13.9 mg casein/g pancreas tissue/min in that of pigs. Differences in the activity of biological raw materials indicate the possibility of studying different options for the preparation of enzyme preparations. A previous study from our lab showed that an injectable drug (lyophilizate from poultry pancreatic tissue) contributed to an increase in the blood pressure of rabbits [8]. The aim of the current study was to determine the effect of crystalline trypsin and lyophilizate from the pancreatic tissue of pigs on the hemodynamic parameters and morpho-biochemical status of rabbits. The current study is also the first study that presents the results of testing a new lyophilic drug prepared from the pancreas of pigs.

2. Materials and Methods

2.1. Animals and Care

The experiments were carried out on rabbits of the Soviet chinchilla breed, age 4.0–6.0 months, body weight of at least 3.5 kg. They were fed with complete granulated feed for rabbits (GOST 32897-2014), in an amount of 100–110 g per day, administered twice per day.

The feed comprised wheat bran, sunflower meal, alfalfa grass meal, barley, wheat, limestone meal, vegetable oil, and a vitamin and mineral premix (crude protein—18.50%, crude fiber—11.13%, metabolizable energy—207.00 Kcal/100 g). Ethical approval was obtained from the Commission on Bioethics of the Russian State Agrarian University—Moscow Timiryazev Agricultural Academy (extract from the protocol application for research using animals No3 from the 7 April 2023 (07/04/02023)). Rabbits in the groups were selected according to the principle of analogs.

The experiment was carried out using the group–period method; the scheme is presented in Table 1.

Table 1. Scheme of the experiment.

Groups			
Group 1 control, n = 5 Injection of intramuscular saline 0.5 mL/head	Group 2, n = 5 Intramuscular injection of trypsin crystalline solution in 0.9% NaCl 0.25 mg/kg body weight	Group 3, n = 5 Intramuscular injection of trypsin crystalline solution in 0.9% NaCl, 0.5 mg/kg body weight	Group 4, n = 5 Intramuscular injection of pig pancreatic lyophilizate 1.0 mg/kg body weight
Period	1 Indicators of hemodynamics and heart rate before injection of the solution		
	2 Hemodynamic and heart rate parameters 60 min after injection of the drug		

2.2. Tested Substances and Their Preparation

Crystalline trypsin (Samson-Med LLC, St. Petersburg, Russia) contains 10 mg of active trypsin per vial, obtained from the pancreas of cattle. Before use, the drug was diluted with saline at a rate of 5.0 mL per vial and injected intramuscularly into the pelvic limb of the rabbit, at a dose of 0.25 mg/kg body weight. When the dose of crystalline trypsin was increased 2-fold, the amount of 0.9% sodium chloride solution was reduced to 2.5 mL and 0.5 mL was injected intramuscularly. Lyophilizate from the pancreas of pigs was prepared using a special technology developed in the scientific laboratory of physiology of animal nutrition of the Russian State Agrarian University—Moscow Timiryazev Agricultural Academy using a lyophilic dryer AK 6–50 (LioGene, Russian Federation). After freezing (−20 °C), the pancreata were placed in vials and dried in a freeze-dryer with the preservation of their properties at the following parameters: temperature −53 °C, pressure 0.015 MPa for 48 h. After drying, the lyophilized powder was packaged in vials of 50 mg and sterilized in an autoclave with saturated steam at 0.5 atm, 114 °C for 30 min. The vials were then sealed with sterile rubber stoppers and aluminum caps. Determination of enzyme activity using the previously described biochemical method [7], using the substrate N-a-Benzoyl-DL-Arginine-4-Nitroanilide Hydrochloride (L-BAPNA) (Sisco Research Laboratories Pvt. Ltd., India), showed that the lyophilized powder has a proteolytic activity of 1768 ± 110.5 U/L [6]. The results of studies to determine trypsin activity in a preparation from the pancreatic tissue of pigs showed that the activity was 432 ± 39.3 U/L. Since the activity of the drugs differed, the dose of lyophilizate chosen for the current study was 1.0 mg/kg of body weight of the rabbits. The drug was diluted before use with 6.25 mL of 0.9% sodium chloride solution and injected suspension intramuscularly into the pelvic limb of the rabbits at 0.5 mL.

2.3. Blood Pressure and Heart Rate Measurements

The procedure for determining the blood pressure (BP) and heart rate in the rabbits was performed daily in the morning hours, before feeding, using an automatic veterinary tonometer—ML-430 VET (Microlux, RF) (Figure 1). To perform this, the rabbits were fixed on a table and the cuff was placed on the front leg of the rabbit and blood pressure was measured at least five times in a row. The device recorded systolic, diastolic, and average blood pressure, as well as heart rate (HR). After injection of the drug, the procedure for measuring blood pressure and heart rate was repeated.

2.4. Blood Sampling

Blood for biochemical studies in rabbits was obtained from the ear vein into test tubes with a coagulation activator containing a silicon oxide (SiO_2) filler. Blood sampling was performed in the morning in a fasted state, after completion of the series of experiments (after 5–7 days). Serum was separated and frozen at −20 °C for further analysis.

2.5. Blood Biochemical Analysis

Blood biochemical analysis (amylase activity, total protein, glucose, triglycerides, cholesterol, alkaline phosphatase, uric acid, calcium, and phosphorus) was performed on an automatic biochemical analyzer—BioChem FC-120 (High Technology, Inc., North Attleboro, MA, USA), using reagent kits from this company. Trypsin activity was determined using a biochemical method on a BS-3000M analyzer (Sinnowa, China), using the substrate N-benzoyl-DL-arginine-n-n-nitroanilide (BAPNA) [7]. For this purpose, 450 µL of buffer solution (pH 8.2)—Reagent 1—was placed into an Eppendorf tube and 50 µL of Reagent 2, containing the trypsin substrate, was then added. Reagent 2 was prepared as follows: 5.0 mg of N-benzoyl-DL-arginine-n-n-nitroanilide (BAPNA) powder was dissolved in 1.0 mL of dimethyl sulfoxide, stirring constantly for 2–3 min. The solution can be stored in a refrigerator at +4 °C for up to 3 months. For the analysis, a mixture of Reagents 1 and 2 was stirred in a closed Eppendorf tube and warmed for 3 min in the thermostat at +37 °C. After incubation, 10.0 µL of the test material (serum) was added and mixed with the reagent mixture and the reaction on the biochemical analyzer was started. The mode for enzyme determination was pre-set on the device: main filter—405 nm, delay—15 s, measurement time—60 s, blank sample—water, factor—4554; calibration was performed using a TruCal multicalibrator (Di Sys Diagnostic Systems GmbH, Germany). The unit of trypsin activity measurement was U/L.

Blood morphological parameters were determined on an automatic hematological analyzer—MicroCC (MicroCC20Plus, MCC-2002-VO-RU, High Technology, Inc., USA).

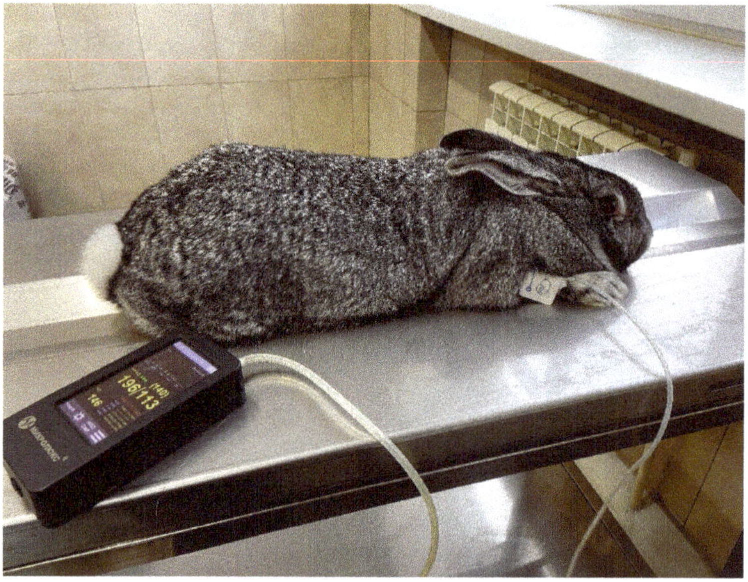

Figure 1. Measuring blood pressure and heart rate in a rabbit.

2.6. Statistical Analysis

STATISTICA 14.0.0.15 (TIBCO Software Inc.) was used for statistical analyses. Data were tested for normal (Gaussian) distribution using the Kolmogorov–Smirnov test, Lilliefors test, or Shapiro–Wilk normality test depending on the sample size. Within-group between-period comparisons were made using either a paired t-test or a Wilcoxon test, depending on the distribution of the data. Differences in investigated parameters between experimental groups were assessed using a one-way ANOVA for normally distributed datasets and a Kruskal–Wallis test for data with non-Gaussian distribution. Homogeneity of variances in the groups was checked with the Levene test. Independent samples with unequal variances were assessed using Welch's ANOVA test. Differences were considered significant if $p < 0.05$ and $p < 0.1$ was considered as a trend.

3. Research Results

3.1. Blood Pressure and Heart Rate

The effects of the pancreatic and crystalline trypsin preparation on the blood pressure and heart rate of rabbits are presented in Table 2.

Table 2. Blood pressure and heart rate in rabbits after intramuscular injections of different enzyme preparations.

Parameter	Group							
	1		2		3		4	
Period	Baseline	Post-injection	Baseline	Post-injection	Baseline	Post-injection	Baseline	Post-injection
Systolic pressure, mmHg	161 ± 3.9	1 62 ± 4.3	167 ± 27.7	148 ± 23.8 *ab	162 ± 32.3	159 ± 31.6 b	166 ± 32.5	152 ± 30.5 *a
Diastolic pressure, mmHg	100 ± 3. 0	99 ± 3.7	104 ± 16.2	92 ± 15.4 *ab	101 ± 20.5	99 ± 20.5 b	104 ± 22.4	96 ± 20.6 *
Mean pressure, mmHg	12 0 ± 2.6	12 0 ± 3.1	125 ± 19.6	111 ± 17.7 *ab	121 ± 24.6	119 ± 23.9 b	125 ± 25.5	114 ± 23.4 *
Heart rate, bpm	215 ± 20.1	212 ± 20.1	226 ± 31.5	210 ± 37.5 *	215 ± 43.2	212 ± 40.4	202 ± 39.4	212 ± 39.9 *

* within-group between-period difference significant at $p < 0.05$; [a]—difference with the control group, [b]—difference between group 2 and group 3 significant at $p < 0.05$. Data are presented as mean ± SD.

In the control group (group 1), no significant changes in blood pressure indices were observed after the injection of 0.9% sodium chloride solution. Table 2 shows that after the administration of crystalline trypsin to rabbits at a dose of 0.25 mg/kg body weight, systolic blood pressure decreased by 11.4% ($p < 0.05$), diastolic blood pressure by 11.5% ($p < 0.05$), mean blood pressure by 11.2% ($p < 0.05$), and heart rate by 7.1% ($p < 0.05$) compared to background.

Interestingly, a 2-fold increase in the dose of crystalline trypsin (group 3) did not result in any changes compared to baseline. At the same time, the injection of pig pancreas lyophilizate (group 4) resulted in a significant decrease in systolic pressure (by 8.4%, $p < 0.05$); diastolic pressure (by 7.7%, $p < 0.05$); and mean arterial pressure (by 8.8%, $p < 0.05$); while heart rate increased by 5.0% ($p < 0.05$). Comparative analysis of the post-injection periods in different groups showed that in group 3, systolic, diastolic, and mean arterial pressure exceeded the values observed in the rabbits of group 2 ($p < 0.05$).

3.2. Blood Biochemical Indices

Blood biochemical indices reflect the metabolic processes and animals' health status. In the current study we determined the activity of digestive enzymes in blood serum to assess the adaptation of the rabbits to the action of the test substances. The results are presented in Table 3.

The intramuscular administration of enzyme preparations had an inhibitory effect on the trypsin activity in the blood of rabbits, which decreased by 51.9% ($p < 0.05$) in the group 3 rabbits compared to the control group (group 1). At the same time, the highest dose of trypsin administered intramuscularly resulted in a significant increase in serum amylase activity by 16.9% ($p < 0.05$) compared to group 1.

An interesting result was obtained with respect to the alkaline phosphatase activity in the serum. This index decreased by 46.4% ($p < 0.05$) in group 2 after the administration of crystalline trypsin at a dose of 0.25 mg/kg body weight, by 43.5% ($p < 0.05$) in group 3 after the administration of crystalline trypsin at a dose of 0.5 mg/kg body weight, but the administration of pig pancreas lyophilizate (group 4) led to a significant increase in alkaline phosphatase activity by 2.2 times ($p < 0.05$) compared to group 1. The level of total protein in the serum also increased slightly in the animals of group 3 (by 7.6%, $p < 0.05$) compared to group 1.

Table 3. Blood biochemical parameters of rabbits after intramuscular injections of different enzyme preparations.

Indicator	Group			
	1	2	3	4
Trypsin activity, U/L	79 ± 34.3	68 ± 2.3	38 ± 12.7 ^{a}	47 ± 9.0
Amylase activity, U/L	154 ± 2.5	156 ± 8.6 ^{b}	180 ± 7.6 ^{abd}	157 ± 7.3 ^{d}
Alkaline phosphatase activity, U/L	69 ± 3.7	37 ± 3.7 ^{ac}	39 ± 10.8 ^{ad}	153 ± 4.3 ^{acd}
Total protein, g/L	66 ± 1.7	65 ± 1.8 ^{b}	71 ± 2.8 ^{abd}	65 ± 1.6 ^{bd}
Uric acid, mmol/L	44 ± 2.6	44 ± 2.0 ^{b}	22 ± 3.0 ^{abd}	39 ± 7.2 ^{d}
Glucose, mmol/L	6.3 ± 0.56	6.0 ± 0.54 ^{c}	6.2 ± 0.74 ^{d}	7.9 ± 1.23 ^{acd}
Triglycerides, mmol/L	0.6 ± 0.06	0.6 ± 0.11	0.6 ± 0.06	0.6 ± 0.04
Cholesterol, mmol/L	2.7 ± 0.59	1.0 ± 0.04 ^{a}	1.0 ± 0.00 ^{a}	1.0 ± 0.00 ^{a}
Calcium, mmol/L	3.3 ± 0.16	3.3 ± 0.04	3.6 ± 0.21	3.4 ± 0.16
Phosphorus, mmol/L	2.9 ± 0.29	1.5 ± 0.05 ^{abc}	0.9 ± 0.16 ^{abd}	3.1 ± 0.33 ^{cd}

^{a}—difference with the control group, ^{b}—difference between group 2 and group 3, ^{c}—difference between group 2 and group 4, ^{d}—difference between group 3 and group 4 significant at $p < 0.05$. Data are presented as mean ± SD.

The serum uric acid levels were significantly reduced in group 3 ($p < 0.05$) compared to the rabbits of group 1. The serum phosphorus level was significantly decreased in the rabbits of group 2 and 3 compared to the rabbits of group 1 ($p < 0.05$).

Morphological blood indices reflect the cellular metabolism and the state of the body's immune system. The results of determining the content of erythrocytes, leukocytes, and platelets in the blood of rabbits are presented in Table 4.

Table 4. Morphological parameters of rabbits during intramuscular injection of different enzyme preparations.

Parameter	Group			
	1	2	3	4
WBC, $\times 10^{9}$/L	6.8 ± 0.46	7.3 ± 0.58 ^{b}	4.8 ± 0.96 ^{abd}	7.6 ± 1.40 ^{d}
LYM, %	28.8 ± 2.17	29.0 ± 2.24 ^{c}	29.5 ± 19.92 ^{d}	47.4 ± 6.31 ^{cd}
GRA, %	65.9 ± 3.37	69.1 ± 2.10 ^{c}	64.6 ± 18.22	48.0 ± 2.92 ^{ac}
Ratio of granulocytes to lymphocytes	2.30 ± 0.26	2.40 ± 0.26 ^{c}	3.2 ± 1.97	1.03 ± 0.19 ^{ac}
RBC, $\times 10^{12}$/L	5.9 ± 0.13	5.9 ± 0.12	6.7 ± 1.47	5.7 ± 0.37
HGB, g/L	135 ± 4.0	137 ± 5.2	133 ± 13.6	142 ± 8.6
MCHC, g/L	350 ± 17.7	343 ± 5.4 ^{bc}	359 ± 7.4 ^{b}	376 ± 12.9 ^{ac}
HCT, %	35.9 ± 3.54	35.9 ± 3.54	41.7 ± 9.88	36.6 ± 0.68
PLT, $\times 10^{9}$/L	491 ± 65.9	558 ± 19.2 ^{b}	430 ± 12.0 ^{bd}	501 ± 30.0 ^{d}
P-LCR,%	8.4 ± 0.69	8.1 ± 0.58 ^{bc}	12.0 ± 2.08 ^{ab}	13.5 ± 2.83 ^{ac}

^{a}—difference with the control group, ^{b}—difference between group 2 and group 3, ^{c}—difference between group 2 and group 4, ^{d}—difference between group 3 and group 4 significant at $p < 0.05$. Data are presented as mean ± SD. WBC—leukocytes count; LYM—lymphocytes count; GRA—granulocytes count; RBC—red blood cells count; HGB—hemoglobin; MCHC—mean corpuscular hemoglobin concentration; HCT—hematocrit; PLT—platelets count; P-LCR—platelets/large cells ratio.

The morphological parameters of the blood of experimental group 2, the rabbits of which were administered crystalline trypsin in the minimum dose (0.25 mg/kg live weight), had no significant differences from control group 1. The number of leukocytes after the administration of the highest dose of crystalline trypsin (group 3) decreased by 29.4% ($p < 0.05$) compared to group 1 and reached the physiological minimum. The administration of pig pancreas lyophilizate did not affect the leukocyte count (group 4). Lymphopenia was observed in control group 1 and experimental groups 2 and 3. Interestingly, the administration of pig pancreas tissue lyophilizate increased the number of lymphocytes by 64.6%, up to the normal reference range. In contrast, the number of granulocytes in the blood of the rabbits of groups 2 and 3 after the administration of enzyme preparations was at the level of group 1, with group 2 being higher than group 4 by 44.0% ($p < 0.05$). At the same time, the ratio of granulocytes to lymphocytes between group 1 and group 4 changed by 1.8 times.

4. Discussion

The current study assessed the effects of various enzyme preparations on the blood pressure, heart rate, and morpho-biochemical blood indices of rabbits. The blood pressure of rabbits has been shown to vary greatly, which becomes relevant when using modern blood pressure measurement devices. Systolic blood pressure in rabbits between the ages of 6 and 8 months has been shown to be 183.5 ± 12 mmHg, whereas diastolic pressure is 125.5 ± 7.5 mmHg. It is known [12,13] that when determining blood pressure in rabbits using an invasive method, the index was 74–100 mmHg. It should be noted that measurements performed by invasive and noninvasive methods may differ.

Enzymes transported into the bloodstream are known to form a pool of deposited zymogenic and active pancreatic enzymes adsorbed by plasma proteins and the endothelium of blood capillaries [14]. These enzymes are desorbed under certain conditions and are included in the pool of enzymes recreated by glandulocytes, together with the pool of enzymes synthesized at the same time by acinocytes. This explains the fact that the amount of pancreatic enzymes transported to the duodenum is substantially greater than the amount that can be synthesized at the maximum of pancreatic secretion, stimulated by food intake or other pathways [15–17]. In our experiments on rabbits, the effects of the enzyme preparations are most probably 'realized' within the nervous system, in the form of a change in blood pressure, during 30–60 min following administration. Previous studies have shown that crystalline trypsin injected into rabbits intramuscularly has an effect on the hemodynamics, heart rate, and morpho-biochemical parameters of the blood. Data presented here indicate that the crystalline trypsin at a dose of 0.25 mg/kg body weight and lyophilizate from the pancreatic glands of pigs at a dose of 1.0 mg/kg body weight had the same effect on the blood pressure of rabbits. Crystalline trypsin at a dose of 0.5 mg/kg body weight had no significant effect on the hemodynamics; apparently, the above dose exceeded the optimal value. Consequently, preparations from the porcine pancreas have an effect on the hemodynamics of rabbits at a certain dose, which suggests the effect of enzymes during parenteral administration on the metabolism of animals. One can assume that the adaptation of rabbits to the action of the enzyme preparations during the five days of the experiment varies depending on the dose and quality of the preparation.

Following the intramuscular injection of trypsin, the trypsin enters the blood and probably inhibits the existing serum trypsin, causing a reaction aimed at changing the metabolism within the whole body. To study the mechanism of action of serum trypsin on animal metabolism, an experiment was conducted to increase the dose of the intramuscular injection of crystalline trypsin 2-fold compared to that used previously. With an increase in the dose of intramuscularly injected trypsin, a significant dose-dependent decrease in serum trypsin activity was noted. For the moment, one thing is obvious—the trypsin preparations increased the influence of the parasympathetic part of the nervous system, which has a positive effect on the assimilation of nutrients and anabolic processes in tissues.

Trypsin also affects hemodynamics through the parasympathetic nerves [18], since the main mechanism of nervous regulation of blood pressure and heart rate is the baroreflex [19], which differs from humoral mechanisms in the speed of reaction. The sensory nerve endings of the depressor nerve, which is a branch of the vagus nerve, are located in the aortic arch. Nerve impulses coming from receptors along the depressor and vagus nerves to the medulla oblongata cause a decrease in the activity of neurons in the pressor zone of the vasomotor center, which leads to an increase in the lumen of blood vessels and a decrease in blood pressure [20–24]. The results of the studies indicate a change in some indicators in the biochemistry of the blood. A decrease in serum trypsin activity with parenteral enzyme preparations may be associated with the regulation of pancreatic secretory function by the feedback principle [25]. It has been reported that a similar response is observed in pancreonecrosis, where exogenous trypsin inhibits trypsin secretion, probably by a short negative feedback mechanism, namely the trypsin–acinar cell axis [26]. The concept of feedback in the regulation of pancreatic secretion emerged as a result of a series of studies demonstrating that the injection of trypsin inhibitor into the upper part of the small intestine or surgical clamping of the bile duct removing bile and pancreatic juice from the duodenum of rats stimulated the secretion of pancreatic enzymes [27–29]. However, the regulation of pancreatic regulation and metabolism by pancreatic enzymes in serum remains poorly understood, and circulating enzyme data are sparse [30]. To date, the mode of entry of pancreatic enzymes into the blood and pancreas remains open to debate [31].

The intramuscular injection of enzyme preparations led to the inhibition of serum trypsin activity. This indicates the competitive effect of trypsin in the blood after intramuscular injection. Changes in trypsin activity in the blood may be due to the fact that trypsin provides a dynamic equilibrium between the activity of proteolytic and inhibitor systems in the process of hemostasis [32] and accelerates the processes of lipid peroxidation (LPO), which is explained by the proteolytic effect of trypsin [33]. An increase in serum trypsin levels is observed in pancreatic inflammation [34,35]. Consequently, despite the data on the effect of trypsin on the regulation of enzyme secretion in the pancreas, the issue of trypsin correlation in the intestine and blood remains poorly studied and requires further research, since the role of trypsin is not limited to protein hydrolysis in the intestine, but goes far beyond this process.

The change in the blood biochemical parameters in the rabbits in group 2 of the current study indicates a change in the metabolism associated with the action of trypsin. The decrease in the trypsin activity in the blood of rabbits of the 3rd and 4th groups, apparently, is associated with the metabolism and adaptation to the intake into the body during the parenteral administration of exogenous trypsin in an increased dose. As the hemodynamic and blood biochemistry parameters show, the application of crystalline trypsin has a more pronounced effect on the organism of rabbits in comparison with the lyophilizate from pig pancreatic gland tissue.

This leads to a change in the ratio of granulocytes to lymphocytes from 2.43 to 1.03, which indicates an increase in the function of the immune system responsible for the specific defense of the body. The number of erythrocytes, which perform the main function of oxygen transport, did not change after the administration of the enzyme preparations.

The ratio of lymphocytes to heterophils in the blood is associated with the state of the immune system, as well as the stress response [36]. It has been established [37] that a change in the average volume of platelets follows a change in the rate of their production. The link between megakaryocyte size and ploidy suggests that lymphoma increases the

DNA content of these cells. Therefore, the data obtained in the current study on the effects of porcine pancreatic lyophilizate and crystalline trypsin on the body of rabbits are new, allowing us to expand the idea of the possibilities of using trypsin as part of complex therapy for circulatory pathology, digestion, and metabolic disorders.

5. Conclusions

The results of the present study of the effect of various preparations containing enzymes of bovine pancreas (crystalline trypsin) and pig pancreas (lyophilizate) showed that after the administration of crystalline trypsin, the mean arterial pressure and heart rate in rabbits decreased by 11.2% ($p < 0.05$), while the administration of pig pancreas lyophilizate decreased the mean arterial pressure by 8.8% ($p < 0.05$). Increasing the dose of crystalline trypsin by 2 times did not lead to a significant change in blood pressure, which indicates that the optimal dose of the drug was exceeded and that there is a regulatory mechanism in the parenteral administration of pancreatic preparations. At the same time, carbohydrate, fat and protein metabolism changed, the number of lymphocytes and the activity of enzymes such as trypsin, amylase and alkaline phosphatase increased, compared to group 1, which was administered with 0.9% NaCl solution. All this indicates the presence of a biologically active substance in the lyophilizate, the effect of which requires further animal studies to create a prototype for the development of new veterinary and medical drugs.

Author Contributions: V.G.V.—conception, design of experiment, text writing; S.V.K.—biochemical and morphological studies of rabbit blood, preparation of substances; N.A.S.—performance of physiological experiments on rabbits; B.S.D.—statistical processing of research materials. All authors have read and agreed to the published version of the manuscript.

Funding: This study did not receive external funding.

Institutional Review Board Statement: All international, national, and institutional principles of animal care and use were observed. All procedures performed in studies involving animals complied with the ethical standards approved by the legislative acts of the Russian Federation, the principles of the Basel Declaration, and the recommendations of the Bioethics Commission of the Institute of Zootechnics and Biology of the Gubkin Russian State Agricultural Academy. K.A. Timiryazev Institute of Animal Science and Biology of the Russian State Agricultural Academy named after K.A. Timiryazev (Minutes No. 7 of 7 April 2023).

Informed Consent Statement: Not applicable.

Data Availability Statement: The original contributions presented in this study are included in the article. Further inquiries can be directed to the corresponding authors.

Conflicts of Interest: There are no conflicts of interest.

References

1. Innerfield, I.; Angrist, A.; Schwarz, A. Parenteral administration of trypsin: Clinical effect in 538 patients. *JAMA* **1953**, *152*, 597–605. [CrossRef] [PubMed]
2. González-Titos, A.; Hernández-Camarero, P.; Barungi, S.; Marchal, J.A.; Kenyon, J.; Perán, M. Trypsinogen and chymotrypsinogen: Potent anti-tumor agents. *Expert Opin. Biol. Ther.* **2021**, *21*, 1609–1621. [CrossRef] [PubMed]
3. Hashem, A.A.R.; Abd El Sattar, A.A.; Abdel Rahman, T.Y. The Effect of Trypsin-Chymotrypsin on Postoperative Pain after Single Visit Endodontic Treatment: A Randomized Controlled Trial. *J. Endod.* **2023**, *49*, 240–247. [CrossRef] [PubMed]
4. Kawabata, A.; Matsunami, M.; Sekiguchi, F. Gastrointestinal roles for proteinase-activated receptors in health and disease. *Rev. Br. J. Pharmacol.* **2008**, *153*, 230–240. [CrossRef] [PubMed]
5. Adams, M.N.; Ramachandran, R.; Yau, M.K.; Suen, J.Y.; Fairlie, D.P.; Hollenberg, M.D.; Hooper, J.D. Structure, function and pathophysiology of protease activated receptors. *Pharmacol. Ther.* **2011**, *130*, 248–282. [CrossRef] [PubMed]

6. Rolland-Fourcade, C.; Denadai-Souza, A.; Cirillo, C.; Lopez, C.; Jaramillo, J.O.; Desormeaux, C.; Cenac, N.; Motta, J.P.; Larauche, M.; Taché, Y.; et al. Epithelial expression and function of trypsin-3 in irritable bowel syndrome. *Gut* **2017**, *66*, 1767–1778. [CrossRef] [PubMed]
7. Koshikawa, N.; Hasegawa, S.; Nagashima, Y.; Mitsuhashi, K.; Tsubota, Y.; Miyata, S.; Miyagi, Y.; Yasumitsu, H.; Miyazaki, K. Expression of trypsin by epithelial cells of various tissues, leukocytes, and neurons in human and mouse. *Am. J. Pathol.* **1998**, *153*, 937–944. [CrossRef] [PubMed]
8. Ramachandran, R.; Hollenberg, M.D. Proteinases and signalling: Pathophysiological and therapeutic implications via PARs and more. *Br. J. Pharmacol.* **2008**, *153*, 263–282. [CrossRef] [PubMed]
9. Kuzmina, I.V.; Ovchinnikova, N.V.; Tolpygo, S.M. Activity of the proteolytic enzyme trypsin in blood serum in rats under conditions of water and food deprivation. *Bull. Exp. Biol. Med.* **2023**, *175*, 540–548. [CrossRef]
10. Vertiprakhov, V.G.; Polina, S.I.; Sergeenkova, N.A.; Sedletskaya, E.S. Hemodynamics and biochemistry of rabbit blood during intramuscular application of trypsin preparations. *Veter-i Korml.* **2024**, *5*, 15–18. [CrossRef]
11. Kuzmina, I.V.; Tolpygo, S.M.; Shoibonov, B.B.; Ovchinnikova, N.V. The influence of soy trypsin inhibitor on biochemical indicators of blood in conditions of water and food deprivation in rats. *Bull. Exp. Biol. Med.* **2024**, *178*, 306–309. [CrossRef]
12. Batoev, T.Z.; Bashanova, I.A.; Koturai, I.A. Enzymatic activity of bovine pancreatic tissue homogenate and its adaptation. *Bull. Buryat State Univ.* **2012**, *4*, 190–192.
13. Vertiprakhov, V.G.; Grozina, A.A. Assessment of the state of the pancreas by determining trypsin activity in poultry blood. *Vet. Sci.* **2018**, *6*, 51–54. [CrossRef]
14. Korotko, G.F. Recirculation of enzymes of digestive glands. In *EDVI*; Editorial House: Krasnodar, Russia, 2011; Volume 144.
15. Gőtze, H.H.; Rothman, S.S. Enteropancreatic circulation of digestive enzyme as a conservation mechanism. *Nature* **1975**, *257*, 607–609. [CrossRef] [PubMed]
16. Heinrich, H.C.; Gabbe, E.E.; Brüggemann, J.; Ičagić, F.; Classen, M. Enteropancreatic circulation of tripsin in man. *Klin. Wschr.* **1979**, *57*, 1295–1297. [CrossRef] [PubMed]
17. Rothman, S.; Liebow, C.; Isenman, L.C. Conservation of digestive enzymes. *Physiol. Rev.* **2002**, *82*, 1–18. [CrossRef] [PubMed]
18. Saidov, S.A. Correlation of body weight gain and blood pressure in modeling metabolic syndrome in rabbits. *BISSA* **2006**, *2*, 53. Available online: https://cyberleninka.ru/article/n/korrelyatsiya-prirosta-massy-tela-i-arterialnogo-rosta-massy-tela-i-arterialnogo-davleniya-pri-modeliro-vanii-metabolicheskogo-sindroma-u-krolikov (accessed on 27 August 2023).
19. Evlakhov, V.I.; Poyasov, I.Z. Changes in pulmonary hemodynamics in experimental myocardial ischemia in rabbits after an increase in arterial pressure. *Russ. Physiol. J. Named After I.M. Sechenov.* **2012**, *3*, 342–351.
20. Vertiprakhov, V.G.; Polina, S.I.; Sedletskaya, E.S.; Sergeenkova, N.A. Hematological and hemodynamic parameters in rabbits (*Oryctolagus cuniculus* subsp. *domesticus*) under the influence of intramuscular injection of trypsin. *Agric. Biol.* **2024**, *59*, 366–374. [CrossRef]
21. Lavanga, M.; Baselli, G.; Fumagalli, F.; Ristagno, G.; Ferrario, M. The possible role of the vagal nervous system in the recovery of the blood pressure control after cardiac arrest: A porcine model study. *Physiol. Meas.* **2017**, *38*, 63–76. [CrossRef] [PubMed]
22. Zhilyaev, S.Y.; Plato-nova, T.F.; Alekseeva, O.S.; Nikitina, E.R.; Demchenko, I.T. Adaptive mechanisms of baroreflex regulation of the cardiovascular system in extreme hyperoxia. *J. Evol. Biochem. Physiol.* **2019**, *55*, 316–323. [CrossRef]
23. Castania, J.A.; Katayama, P.L.; Brognara, F.; Moraes, D.J.A.; Sabino, J.P.J.; Salgado, H.C. Selective denervation of the aortic and carotid baroreceptors in rats. *Exp. Physiol.* **2019**, *104*, 1335–1342. [CrossRef] [PubMed]
24. Dergacheva, O. Chronic intermittent hypoxia alters neurotransmission from lateral paragigantocellular nucleus to parasympathetic cardiac neurons in the brain stem. *J. Neurophysiol.* **2015**, *113*, 380–389. [CrossRef] [PubMed]
25. Sheverdin, Y.P. Inhibition of endogenous insulin and trypsin secretion by exogenous insulin and trypsin in the complex treatment of acute pancreatitis. *Vest. Surg.* **1989**, *1*, 54–55.
26. Adeghate, E.; Ember, Z.; Donath, T.; Pallot, D.J.; Singh, J. Immunohistochemical identification and effects of atrial natriuretic peptide, pancreastatin, leucine-enkephalin, and galanin in the porcine pancreas. *Peptides* **1996**, *17*, 503–509. [CrossRef] [PubMed]
27. Adler, G. Effect of glucagon on the secretory process in the rat exocrine pancreas. *Cell Tissue Res.* **1977**, *182*, 193–204. [CrossRef] [PubMed]
28. Green, G.M.; Lyman, R.L. Feedback regulation of pancreatic enzyme secretion as a mechanism for trypsin inhibitor-induced hypersecretion in rats. *Proc. Soc. Exp. Biol. Med.* **1972**, *140*, 6–12. [CrossRef] [PubMed]
29. Lokshina, L.A. Regulatory role of proteolytic enzymes. *Molecules. Biol.* **1979**, *13*, 1205–1229.
30. Leibow, C.; Rothman, S.S. Enteropancreatic circulation of digestive enzymes. *Science* **1975**, *189*, 472–474. [CrossRef] [PubMed]
31. Levitt, M.D.; Ellis, C.J.; Murphy, S.M.; Schwartz, M.L. Study of the possible enteropancreatic circulation of pancreatic amylase in the dog. *Am. J. Physiol.* **1981**, *241*, G54–G58. [CrossRef] [PubMed]
32. Nasyrov, K.H.M.; Kondratenko, R.M. On the prooxidant action of inflammatory mediators. *Stalemate. Physiol. Exp. Ter.* **1992**, *3*, 12–14.
33. Kalagina, L.S. Clinical value of determination of trypsin indices in blood serum (review). *Med. Almanac.* **2010**, *4*, 281–283.

34. Degtyareva, I.I.; Gaisenko, A.V.; Putseva, N.M. Laboratory diagnosis of pancreatitis and pancreatic cancer. *Dr. Bus.* **1989**, *7*, 42–43.
35. Fushiki, T.; Iwai, K. Two hypotheses on the feedback regulation of pancreatic enzyme secretion. *FASEB J.* **1989**, *3*, 121–126. [CrossRef] [PubMed]
36. Scanes, C.G. Biology of stress in poultry with emphasis on glucocorticoids and the heterophil to lymphocyte ratio. *Poult. Sci.* **2016**, *95*, 2208–2215. [CrossRef] [PubMed]
37. Gladwin, A.M.; Martin, J.F. The control of megakaryocyte ploidy and platelet production: Biology and pathology. *Int. J. Cell Cloning* **1990**, *8*, 291–298. [CrossRef] [PubMed]

biomolecules

MDPI

Article

The Intestinal Mechanisms in the Excretion of Pepsinogen, Amylase and Lipase in Coprofiltrate in Women During Pregnancy and the Postpartum Period

Elena Kolodkina and Sergey Lytaev *

Department of Normal Physiology, St. Petersburg State Pediatric Medical University, 194100 Saint Petersburg, Russia; 922-666-2045@mail.ru
* Correspondence: physiology@gpmu.org; Tel.: +7-812-416-5226

Abstract: Background: Enzymes secreted by the digestive glands are excreted from the body with urine, sweat and feces, and they are also removed from the blood due to their participation in the enzymatic provision of the secretion entering the gastrointestinal tract. **Objective**: The aim of this work was to analyze the activity of pepsinogen, amylase and lipase in the coprofiltrate of pregnant women in each trimester of pregnancy and in the postpartum period, taking into account the timing and type of delivery (term, premature, late delivery or cesarean section). **Methods**: Data from studies of non-pregnant (n = 45) and pregnant (n = 193) women were analyzed. The materials for preparation coprofiltrate were collected during delivery. Pepsinogen activity was determined by proteolytic activity at pH = 1.5–2.0 using the tyrosine spectrophotometric method, while amylase activity was determined by the amyloclastic method of Karavey, and lipolytic activity was determined by a unified kinetic method using olive oil as a substrate. **Outcomes**: A small amount of pepsinogen was excreted in the coprofiltrate, and while the level of its excretion increased after childbirth, it remained below the control values. At the same time, an increase in the amylolytic activity of the coprofiltrate was observed in all groups of pregnant women examined from the first to the third trimester of pregnancy. In pregnant women, multidirectional changes in lipase activity were observed depending on the timing and type of delivery. **Conclusions**: At the end of pregnancy, amylolytic activity increased in all women, and pepsinase activity decreased compared to the indicators of non-pregnant women. No reliable differences were found in the lipolytic activity of the coprofiltrate in pregnant women at the end of pregnancy and the indicators of non-pregnant women.

Keywords: pregnancy; digestive enzyme; coprofiltrate; mother-fetus functional system

1. Introduction

Hydrolases enter the digestive tract from the blood, which are then secreted into the intestine and participate in the processing of digestive substrates through step-by-step cavity, peri-mucosal and membrane digestion [1,2]. The restoration of digestive enzymes in the gastrointestinal tract is clearly expressed. Experimental studies on animals, in particular on dogs, have shown that isolated intestinal loops for 20–25% of the digestive capacity are achieved without the participation of pancreatic or prepancreatic digestive enzymes [3,4].

The mechanisms of incretion of digestive gland enzymes have been studied previously [4,5]. The problem of the biological significance of blood hydrolase homeostasis has been partially resolved, and their diverse role in the body has been revealed, including the anabolic, regulatory, informational, transport and other functions of pepsinogen, amylase, and lipase [6–8]. According to the data obtained, the concentration of serum amylase reflects the balance between the rates of amylase entry into the blood and its removal from it. Hyperamylasemia can result from either an increased rate of amylase entry into the bloodstream or a decreased metabolic clearance of this enzyme [9–12].

Physiological changes associated with pregnancy are the result of hormonal and metabolic changes necessary to support the growing fetus. In addition, they lead to changes in the functioning of many body systems, including the gastrointestinal tract [13]. In particular, the effect of pregnancy on irritable bowel syndrome has been shown alongside the efficacy and safety profiles of widely used diets and drugs for this syndrome during pregnancy [14].

During pregnancy, bilateral metabolic relationships are formed between the mother's organism and the developing fetus [15,16]. The fetus absorbs some of the nutrients by swallowing amniotic fluid, which are enzymatically hydrolyzed to monomers in the gastrointestinal tract of the developing organism. In this regard, digestive hydrolases are of particular importance in the functional system "maternal body—placenta, amniotic fluid—fetus" [17–19].

Pregnancy affects homeostasis mechanisms involving the intestine, which is associated with disturbances in the mother–fetus system [20–24]. Secreted enzymes participate in cavity and parietal hydrolysis [4–6]. Participation in this digestion of exocrine enzymes, which have not been degraded after their secretion in the overlying sections of the gastrointestinal tract, is possible [25]. In the small intestine, under normal conditions, chyme is formed [25], which is a nutrient medium for eubiotic symbionts of the large intestine, a violation of the composition of which leads to dysbiosis [26]. Along with symbiotic digestion with the participation of enzymes of the bacterial flora, the further hydrolysis of undigested products occurs here by enzymes of resecretory and exocrine origin [27,28]. In the absence of a microvillous apparatus in the large intestine, bacterial cells serve as structures that absorb enzymes [29].

A number of authors have studied changes in the composition of the intestinal microbiome that occur between the first and third trimesters of pregnancy. In particular, an increase in the number of Akkermansia, Bifidobacterium and Firmicutes has been noted, which is associated with an increase in the need for energy accumulation. An increase in the levels of Proteobacteria and Actinobacteria has a protective effect on the mother and the fetus [30].

There is a need to monitor gastrointestinal diseases during pregnancy; in this regard, expert recommendations are proposed regarding the clinical management of patients with gastrointestinal and liver diseases associated with pregnancy [20], determining the use of pharmacokinetic variables of drugs during pregnancy through taking into account the enzymatic activity of biological fluids [21].

Thus, unused hydrolases are excreted in the feces during defecation, and their hydrolytic activity can be detected in the coprofiltrate. Based on the above, the present study was aimed at studying the features of the mechanisms of enzymatic activity of pepsinogen, amylase and lipase in the coprofiltrate of pregnant and parturient women during the trimesters of pregnancy and in the postpartum period, depending on the type and timing of delivery (full-term, premature or late delivery, or cesarean section).

2. Materials and Methods

This study involved 193 Caucasian, primiparous pregnant women aged 18 to 35 years with a physiological course in the gestation period. They did not have any pathology of the gastrointestinal tract or liver and did not use drugs, alcohol, or nicotine. To select patients, a clinical interview and preliminary questionnaire of pregnant women were conducted to exclude pathologies of organs and systems. The first group consisted of 161 pregnant women whose births ended through the natural birth canal (96 with term births, 34 with premature births, and 31 with late births—more than 40 weeks). The second group consisted of 32 pregnant women whose births ended in emergency surgery (cesarean section). For select patients, a clinical interview and preliminary questionnaire of pregnant women were performed to exclude pathology of the digestive system and liver. Screening for hepatitis viruses is mandatory in accordance with the national standard. During the study period, no participants reported any complaints of diarrhea or constipation. These data were constantly clarified during the interview with the participants.

The gestational age at the time of delivery was determined by the date of last menstruation, the date of the first fetal movement, the first visit to the antenatal clinic, and data from external obstetric and ultrasound examinations.

The clinical data on the course of pregnancy and childbirth were obtained based on the exchange and notification card for the observation of the pregnant woman and the woman in labor, the history of childbirth, and the history of the development of the newborn.

The control group consisted of 45 practically healthy non-pregnant women aged 18 to 30 years (average 24.2 $\pm$ 0.3 years) without any pathologies or concomitant diseases.

This study on women in the St. Petersburg Snegirev's Maternity Hospital was performed. All women (the control and research groups) were informed about the purpose and methods and gave written voluntary informed consent to participate in the study (protocol no. 0608-23 dated 7 August 2023, of the Local Ethics Committee of the Almazov National Medical Research Center).

The research design consisted of the following stages: coprofiltrate was tested once in healthy subjects (women) and four times in pregnant women regardless of the type of delivery in each trimester of pregnancy and after the birth of the child. To prepare the coprofiltrate, 1 g of the feces being examined was mixed with 10 mL of distilled water. The resulting mixture was then passed through a paper filter. The coprofiltrate was collected for enzyme content determination at the same time of day under the same conditions (in the morning, on an empty stomach). The coprofiltrate was collected once from healthy subjects (women) and four times from pregnant women in each trimester of pregnancy and after childbirth. Early coprofiltrate is a stool sample used to diagnose various conditions related to the gut in the early stages of a disease or condition. Such studies may include an analysis of gut microflora, enzymatic activity, and other parameters. Coprofiltrate collection for enzyme content determination was performed at the same time of day under the same conditions (morning, fasting). Coprofiltrate was collected once from healthy subjects (women) and four times from pregnant women, regardless of the type of disease, in each trimester of pregnancy and after childbirth.

The activities of pepsinogen, amylase and lipase were determined in the coprofiltrate of pregnant women at 12–13, 25–26, and 39–40 weeks of pregnancy and once in individuals in the control group.

Pepsinogen activity was determined by proteolytic activity at pH = 1.5–2.0 by the tyrosine spectrophotometric method. The method is based on the fact that tyrosine, which is formed during the hydrolysis of pepsinogen, absorbs ultraviolet light at a wavelength of about 280 nm. This allows one to determine the concentration of tyrosine and, accordingly, the activity of pepsinogen. The principle of the method is that the protein substrate is incubated with the test liquid (coprofiltrate). The measure of its activity is the amount of tyrosine formed as a result of protein hydrolysis. In this study, lyophilized blood plasma was used as a protein substrate.

Amylase activity was determined using the Karavey amyloclastic method [2]. Experimental and control samples were taken for analysis. A starch solution was added to the experimental sample, which was followed by heating for 5 min in a water bath at 37 °C and adding 0.02 mL of non-hemolyzed serum. The sample was again placed in a water bath for 5 min. The control sample was carried out in the same way but without serum. After that, 1 mL of the working iodine solution and 8 mL of distilled water were added to all test tubes (experimental and control). The solutions were colorimetrically measured on a photometer with a red filter (630–690 nm) in a cuvette with a layer width of 10 mm. Distilled water was used as a standard. A reagent kit for determining the activity of alpha-amylase "Alpha-Amylase-1-Olvex", Olvex Diagnosticum LLC (Saint-Petersburg, Russia) was used.

Lipolytic activity was determined by a unified kinetic method using olive oil as a substrate, where the unit of enzymatic activity is the amount of enzyme that releases 1 μmol of oleic acid from a 40% olive oil emulsion at pH 7.0 and 37 °C for 1 h. The principle of the method is to spectrophotometrically measure the turbidity of an olive oil suspension under the action of an enzyme. A reagent kit for the quantitative determination of lipase activity by the kinetic method "Lipase-UTS", Eiliton LLC (Dubna, Russia) was used.

All methods used to study enzymatic activity are approved by national standards and comply with GLP standards.

Statistical Analysis

The enzyme levels in non-pregnant women were taken as controls to show the differences from those in pregnant women. First, the distribution was checked for normality using the Pearson criterion. The average values (M) and their standard error (m) and standard deviation (σ) were calculated. The dependence between the features was assessed using the pair correlation coefficient (r), its error (mr), and the level of significance of differences (according to Student's t-criterion). The dependence was considered strong when $|r| > 0.7$ (average) if the modulus of the pair correlation value was within 0.3–0.7. A correlation value of less than the modal value of 0.3 was considered weak.

The non-parametric Mann–Whitney criterion was used when a non-normal distribution was confirmed. Differences between groups in the levels of the studied characteristics were estimated with the non-parametric Mann–Whitney method using the statistical package SPSS 11.0. Differences were considered significant at an error probability of $p < 0.05$. Significance levels of $p < 0.01$ and $p < 0.001$ were also distinguished. The results are presented as Me [Q1; Q3], where Me is the median, while Q1 and Q3 are the lower and upper quartiles, respectively. The results were statistically processed using Microsoft Excel 2003 spreadsheets and the SPSS 23.0 and Primer of Biostatistics 4.03 programs.

3. Results

We have found that out of 82 pregnant women examined during the year, 64 women (78.1%) gave birth through the natural birth canal (42 pregnant women with full-term birth, 12 pregnant women with premature birth, 10 pregnant women with late birth), and 18 women (21.9%) gave birth by emergency surgery—cesarean section.

The average age of pregnant women with natural birth was 30.6 ± 2.1 years (with full-term birth −29.1 ± 1.9 years, premature birth −30.2 ± 3.4 years, late birth −30.8 ± 2.7 years), and in women who gave birth by cesarean section, it was −32.3 ± 5.9 years.

Clinical data in pregnant women with different terms and types of delivery (term delivery (group 1), premature delivery (group 2), late delivery (group 3), delivery by cesarean section (group 4)) are presented in Table 1.

Table 1. Comparative characteristics of clinical indicators in pregnant women with natural delivery and by caesarean section, *—$p < 0.05$, **—$p < 0.001$.

Parameters	Groups of Pregnant Women			
	Timely Birth (n = 86)	Premature Birth (n = 34)	Late Birth (n = 31)	Childbirth by Cesarean Section (n = 42)
Pre-gestational body weight, kg	62.4 ± 1.2	53.5 ± 1.6 *	64.8 ± 1.4 **	60.1 ± 2.2
Height, cm	164.2 ± 12.5	164.6 ± 14.8	168.5 ± 12.7	166.8 ± 13.2
Total weight gain during pregnancy, kg	9.3 ± 0.2	10.3 ± 0.8	12.6 ± 0.7 *	11.8 ± 0.9 **
Pre-gestational body mass index, kg/m^2	22.0 ± 1.4	20.4 ± 1.8 **	24.5 ± 2.1 **	22.8 ± 2.0
Early toxocosis, %	2.8	2.2	2.5	3.1
Anemia, %	1.5	1.6	1.3	1.4
Weakness of labor activity, %	4.3	3.7	4.8	5.3

Pepsinogen activity in the coprofiltrate (dilution 1:4) in pregnant women who completed their pregnancy with delivery at term was low and amounted to 410.2 ± 34.1 thousand units/mL. It decreased throughout pregnancy and in the third trimester was ¼ of the initial level −153.8 ± 11.2 thousand units/mL. On the 2nd–3rd day after delivery, the pepsinase activity of the coprofiltrate increased by 2.1 times ($p < 0.001$) compared to the end of pregnancy, but it remained lower −315.3 ± 23.6 thousand units/mL ($p < 0.05$) compared to the control group (Figure 1).

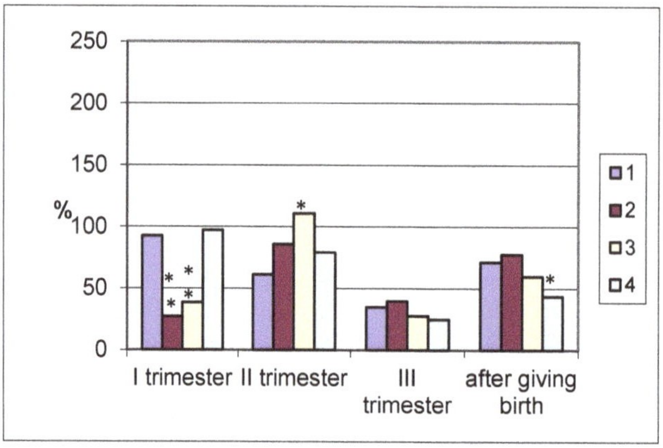

Figure 1. Pepsinogen activity in the coprofiltrate of pregnant women (by trimester of pregnancy and after delivery) with different timing and types of delivery depending on the timing and type of delivery (in % of the control group values taken as 100%). Note. 1—term delivery; 2—premature delivery; 3—late delivery; 4—delivery by cesarean section; significance of differences with values in pregnant women who delivered at term: *—$p < 0.05$; **—$p < 0.001$.

In pregnant women with premature births, there was a decrease in the proteolytic activity of the coprofiltrate throughout pregnancy and in the postpartum period compared to the control values. In the first trimester of pregnancy, pepsinogen activity was 3.5 times lower ($p < 0.001$) than that in non-pregnant women, while it was 1.1 times lower in the second trimester, 2.4 times lower in the third ($p < 0.001$), and 1.2 times lower after childbirth($p < 0.05$).

Different dynamics were observed in relation to pepsinogen in the coprofiltrate of pregnant women with late labor. There was a decrease in the enzyme activity in the first trimester of pregnancy −172.0 ± 11.3 tyr. U/mL ($p < 0.001$) compared with the control values, while there was an increase in the second trimester −493.1 ± 24.8 tyr. U/mL and decreases in the third trimester (122.4 ± 10.3 tyr. U/mL ($p < 0.05$) and in the postpartum period (263.8 ± 13.7 tyr. U/mL ($p < 0.001$). Such losses of pepsinogen may indicate a greater use of the enzyme in the anabolism of substances in the mother–fetus system, and it may also reflect a decrease in incretion during pregnancy.

Pepsinogen excretion in feces of pregnant women with cesarean section in the first trimester was almost equal to the values of non-pregnant women and amounted to 430.4 ± 32.7 tyr. U/mL (in the control 422.2 ± 28.4 tyr. U/mL). In the second and third trimesters, the pepsinase activity of the coprofiltrate decreased, with the greatest changes at the end of pregnancy −110.3 ± 11.8 tyr. U/mL ($p < 0.001$), remaining low in the postpartum period (192.9 ± 10.6 tyr. U/mL; $p < 0.001$).

In the postpartum period, pregnant women with urgent, premature, late labor, and cesarean section delivery showed a tendency toward increased intestinal enzyme secretion, not reaching the 100% levels seen in non-pregnant women. Thus, a small amount of pepsinogen is secreted in the coprofiltrate, and only after delivery does its secretion level increase, but it remains below control levels. In the postpartum period, intestinal enzyme secretion is maintained for use in subsequent interactions between the nursing mother and her child.

The amylolytic activity of the coprofiltrate in the group of pregnant women who gave birth at term increased by more than 50% in the second and by 127% in the third trimester, decreasing to the initial values after delivery (Figure 2). In pregnancies ending in premature delivery, similar dynamics of changes in the amylolytic activity of the coprofiltrate are observed, as in the case of term delivery, with a slight difference: a reduced enzyme content in the coprofiltrate.

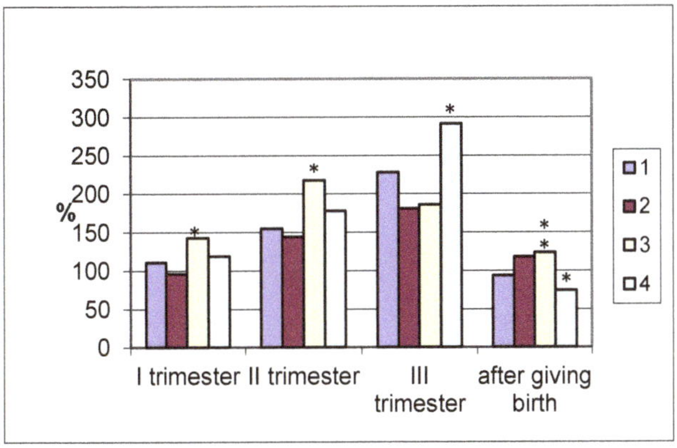

Figure 2. Coprofiltrate amylase activity in pregnant women (by trimester of pregnancy and after delivery) with different timing and types of delivery depending on the timing and type of delivery (in % of the control group values taken as 100%). Note: 1—term delivery; 2—premature delivery; 3—late delivery; 4—delivery by cesarean section; significance of differences with values in pregnant women who delivered at term: *—$p < 0.05$; **—$p < 0.001$.

In pregnant women with late labor, the excretion of amylase with feces exceeded the coprofiltrate indices of non-pregnant women. In the first and second trimesters of pregnancy and in the postpartum period, the enzyme activity in the coprofiltrate of the studied group of women was the highest compared to other groups.

In women with cesarean section, the amylolytic activity of the coprofiltrate increased by trimester of pregnancy. In the first trimester, it was 118.9% that of the control, while in the second trimester, it was 177.9%, and in the third trimester, it was 291.2%. In the postpartum period, it sharply decreased, amounting to 74.3%, which is lower than the indicators of non-pregnant women. Thus, an increase in the amylolytic activity of the coprofiltrate was observed in all groups of pregnant women examined, grouped by the timing and types of delivery, from the first to the third trimester of pregnancy.

Lipase excretion by the intestines differed significantly in different groups of women studied. In pregnant women who gave birth at term, the enzyme activity in the coprofiltrate had the highest values in the first trimester (420.3 ± 22.5 U/mL; ($p < 0.001$), which is 31% higher than in non-pregnant women (Figure 3). Subsequently, during pregnancy, it decreased and after delivery was the same as in the control.

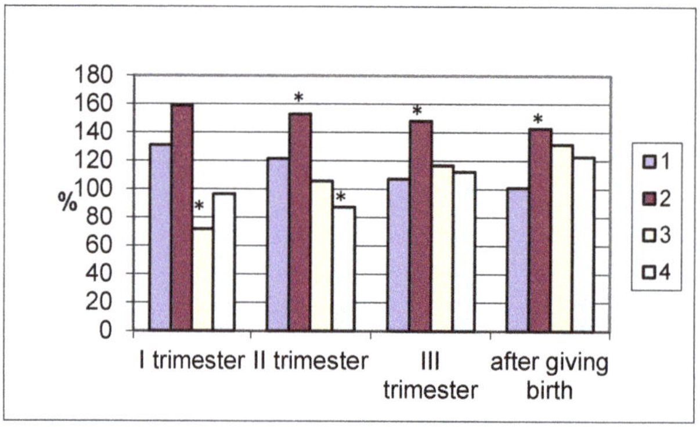

Figure 3. Coprofiltrate lipase activity in pregnant women (by trimester of pregnancy and after delivery) with different timing and types of delivery depending on the timing and type of delivery (in % of the control group values taken as 100%). Note: 1—term delivery; 2—premature delivery; 3—late delivery; 4—delivery by cesarean section; significance of differences with values in pregnant women who delivered at term: *—$p < 0.05$.

Similar dynamics were observed in women with premature delivery (maximum lipase values in the coprofiltrate were in the first trimester, while the lowest were in the postpartum period). A different pattern of changes in the lipolytic activity of the coprofiltrate was observed in women with late delivery. Being reduced in the first trimester of pregnancy compared to the control group data, the excretion of the enzyme in the coprofiltrate increased in the second (105.9%) and third (116.9%) trimesters, reaching maximum values after delivery (131.6%). In women who delivered by cesarean section, during pregnancy and after delivery, differently directed dynamics of changes in intestinal lipase excretion were observed. In the first trimester of pregnancy, the enzyme activity in the coprofiltrate was practically no different from the control group values and was 96.6% of the control. In the second trimester, the lipolytic activity of the coprofiltrate decreased and was 12.7% lower than in non-pregnant women. In the third trimester of pregnancy and after childbirth, intestinal lipase excretion increased, exceeding the control group values by 12.4% and 22.8%.

4. Discussion

Thus, pregnancy affects homeostasis mechanisms involving the intestine, which is associated with hormonal and metabolic shifts in the "mother–fetus" system as well as with disruption of the biorhythms of bowel movements. Various mechanisms participate in ensuring the homeostasis of pepsinogen, amylase and lipase, among which is the excretion of hydrolases of the digestive glands [3,5]. Pepsinogen, amylase, and lipase enter the digestive tract; then, these enzymes are resected into the intestine and participate in the complexly organized digestion of substrates through step-by-step cavity, primucosal and membrane digestion [2,4,5].

Pepsinogen is expressed in the main cells of the stomach; the fundic mucous cells and pyloric glandular cells of the gastric epithelium; the mammary gland, prostate gland, lungs and seminal vesicles; and changes in pathological conditions of the body as well as during pregnancy [31].

The results of the present study show that women with term delivery during pregnancy have low pepsinogen activity compared to the control group. This fact toward the end of pregnancy can be associated with the expenditure or consumption of enzymes on anabolic processes of the growing fetus. After birth, the enzyme level increased, but the activity remained lower than in non-pregnant women. The same dynamics were observed in women with premature births, while in late births, there were multidirectional changes in the enzyme activity in the coprofiltrate during pregnancy and in the postpartum period. In women with cesarean section in the first trimester of pregnancy, there were no differences in enzyme activity from the control group, in the second, third trimester and after birth, there was a decrease in the proteolytic activity of the coprofiltrate.

The amylolytic activity of blood serum is normally realized by almost half by the action of pancreatic α-amylase [5]. As is known, amylases have a number of phenotypes that differ in their physicochemical properties, and they also have different renal clearance and an unequal half-life from the body [32]. In particular, there is information on the genetic characteristics of the ratio of β- and s-amylases in human blood serum [33]. Amylases in the blood are free and bound to plasma proteins (serum) and formed elements. At the same time, plasma proteins (albumins and all globulin fractions) have independent amylolytic activity. This is one of the forms of enzyme deposition: to maintain homeostasis of the body's enzymes by increasing or decreasing the affinity for them depending on hypo- or hyperamylasemia [34].

The amylolytic activity of coprofiltrate in pregnant women with term births and cesarean section increased from the beginning to the end of pregnancy and decreased in the postpartum period. In case of premature birth, the enzyme activity in the coprofiltrate in the first trimester of pregnancy did not differ from the control group indicators, and in the second and third trimesters as well as after birth, it was higher than the control group indicators. In case of late birth, there was an increase in amylase activity in the coprofiltrate throughout pregnancy and in the postpartum period.

The origin of lipase in blood serum is mainly pancreatic and hepatic. It is also found in formed elements of blood (erythrocytes, leukocytes), but serum and erythrocyte lipolytic activities are independent of each other [35].

In the small intestine, membrane digestion is carried out on the microvilli of enterocytes coupled with the processes of transport of nutrient monomers into the villus. In the large intestine, the removal of enzymes from the cavity can be carried out by microorganisms—saprophytes participating in symbiotic digestion.

Unused hydrolases are excreted in the feces during defecation, and their hydrolytic activity can be detected in the coprofiltrate prepared from the feces by dilution with saline in a ratio of 1:4.

The lipolytic activity of coprofiltrate in pregnant women in our study with term and premature births increased in the first trimester, with a subsequent decrease in the second and third trimester as well as after birth. The opposite dynamics was noted in pregnant women with late births—the lowest values were in the first trimester, and the highest were after birth. Multidirectional changes in lipase activity were observed in women with cesarean section (no changes in the first trimester compared to the control, a decrease in the second trimester, an increase in the third trimester and after birth).

During pregnancy, metabolic connections are established between the mother's body and the growing fetus, which actively absorbs amino acids for protein synthesis. The fetus absorbs nutrients with amniotic fluid, which are hydrolyzed to monomers in the gastrointestinal tract of the developing organism by enzymes secreted into the aquafetal environment during autolytic digestion [35–37].

As a result of our studies, the excretory origin of the hydrolytic activity of the coprofiltrate was revealed due to the detection of amylase, pepsinogen and lipase in feces. At the end of pregnancy, amylolytic activity increased in all women, while pepsinase activity decreased by 3 times ($p < 0.001$) compared to the indicators of non-pregnant women. No reliable differences were found in the lipolytic activity of the coprofiltrate in pregnant women at the end of pregnancy and the indicators of non-pregnant women.

Thus, as a result of our studies, it was revealed that at the end of pregnancy, amylolytic activity increased in all women, while pepsinase activity decreased compared to the indices of non-pregnant women. No reliable differences were found in the lipolytic activity of coprofiltrate in pregnant women at the end of pregnancy and the indices of non-pregnant women. During pregnancy, neuro-endocrine changes occur in the female body, while the important role belongs to hormonal shifts. There are intense metabolic processes between the mother's body, placenta and the fetus. The placenta begins to produce hormones. Changes in the functioning of the liver and pancreas are noted. All these factors affect the change in enzymes during pregnancy in women.

5. Limitations

The study included pregnant women with a normal gestation period, primiparous women, and those without gastrointestinal tract pathology. Multiparous women were not taken into account. This requires further research with a wider range of pregnant women. The study examines the activity of digestive enzymes secreted by the intestine during pregnancy and after childbirth. The level of enzymes in other biological fluids (gastric juice) was not assessed. This requires further research with a wider range of assessment of the gastrointestinal tract in pregnant women and nursing mothers. All subjects underwent diagnostic procedures in accordance with current standards—blood tests, necessary functional diagnostics, etc. However, this information is not provided in this paper, since it is not the purpose of this study; instead, they may be presented later.

6. Conclusions

The results obtained in the present study point to a number of mechanisms of hydrolase homeostasis: excretion from the blood by the intestine, the action of digestive juices, and the restoration and removal of enzymes. In particular, a small amount of pepsinogen is eliminated in the feces; after delivery, its level of elimination increases but remains below control values. In the postpartum period, pepsinogen homeostasis in the intestine is maintained for its use in the relationship between mother and child care.

The amylolytic activity of coprofiltrate increases in all groups of pregnant women, according to the timing and type of delivery, from the first to the third trimester of pregnancy. Alternatively, this task can be saved for future studies—studying coprofiltrate amylase in pregnant women without gestational diabetes in comparison with pregnant women with GDM. These data may provide a non-invasive way to screen people at risk of gestational diabetes.

For intestinal lipase removal, the heterogeneous dynamics of changes in the indices of women who gave birth by cesarean section during pregnancy and after childbirth are noted. In the first trimester of pregnancy, lipase activity is practically no different from the control group. In the second trimester, the lipolytic activity of the coprofiltrate decreases compared to non-pregnant women. In the third trimester of pregnancy and after childbirth, intestinal lipase excretion increases, significantly exceeding the indices of the control group.

Author Contributions: Conceptualization, E.K. and S.L.; methodology, E.K.; software, E.K.; validation, E.K. and S.L.; formal analysis, E.K.; investigation, E.K.; resources, E.K.; data curation, E.K.; writing—original draft preparation, E.K.; writing—review and editing, S.L.; visualization, S.L.; supervision, S.L.; project administration, S.L.; funding acquisition, S.L. All authors have read and agreed to the published version of the manuscript.

Funding: This research by Saint Petersburg State Pediatric Medical University was supported. Project No. AAAA-A19-119112290090-9.

Institutional Review Board Statement: This study was conducted according to the Declaration of Helsinki and approved by the Local Ethics Committee of the Almazov National Medical Research Center (protocol no. 0608-23; approved on 7 August 2023).

Informed Consent Statement: Informed consent was obtained from all subjects involved in this study.

Data Availability Statement: The data for this project are confidential but may be obtained under the data use agreements of the Saint Petersburg State Pediatric Medical University, the Head of the Department of Normal Physiology. Researchers interested in accessing the data may contact Sergey Lytaev at physiology@gpmu.org. It could take some weeks (months) to negotiate the data use agreements and gain access to the data. The author will assist with any reasonable replication attempts for one year following publication.

Conflicts of Interest: The authors declare no conflicts of interest.

References

1. Pediatric gastroenterology. In *National Textbook*; Belmer, S.V., Razumovsky, A.Y., Khavkin, A.I., Eds.; GEOTAR-Media: Moscow, Russia, 2024.
2. Litwack, G. *Human Biochemistry*; Academic Press: Cambridge, MA, USA, 2018.
3. Korot'ko, G.F. Recretion of enzymes and hormones by exocrine glands. *Uspekhi Fiziol. Nauk* **2003**, *34*, 21–32. [PubMed]
4. Korot'ko, G.F. Exo- and endosecretive digestive glands of enzymes as modulators of secretion. *Eksperimental'naia Klin. Gastroenterol.* **2010**, *12*, 81–86. [PubMed]
5. Korot'ko, G.F. Recirculation and adaptation of enzymes of digestive glands secretion. *Exp. Clin. Gastroent.* **2022**, *4*, 138–152. [CrossRef]
6. Timofeeva, N.M. Early metabolic/nutritional programming of enzyme systems of digestive and non-digestive organs. *Gastroenterol. St. Petersburg* **2014**, *1*, 38–40.
7. Pieper-Bigelow, C.; Strocchi, A.; Levitt, M.D. Where does serum amylase come from and where does it go? *Gastroenterol. Clin. N. Am.* **1990**, *19*, 793–810. [CrossRef]
8. Ko, J.; Cho, J.; Petrov, M.S. Low serum amylase, lipase, and trypsin as biomarkers of metabolic disorders: A systematic review and meta-analysis. *Diabetes Res. Clin. Pract.* **2020**, *159*, 107974. [CrossRef]
9. Nakajima, K. Low serum amylase and obesity, diabetes and metabolic syndrome: A novel interpretation. *World J. Diabetes* **2016**, *7*, 112–121. [CrossRef]
10. Qin, Y.; Geng, J.X.; Huang, B. Clinical value of serum pepsinogen in the diagnosis and treatment of gastric diseases. *World J. Gastrointest. Oncol.* **2023**, *15*, 1174–1181. [CrossRef]
11. Lee, S.Y. Endoscopic gastritis, serum pepsinogen assay, and Helicobacter pylori infection. *Korean J. Intern. Med.* **2016**, *31*, 835–844. [CrossRef]
12. Massarrat, S.; Haj-Sheykholeslami, A. Increased Serum Pepsinogen II Level as a Marker of Pangastritis and Corpus-Predominant Gastritis in Gastric Cancer Prevention. *Arch. Iran. Med.* **2016**, *19*, 137–140.
13. Chandra, M.; Paray, A.A. Natural Physiological Changes during Pregnancy. *Yale J. Biol. Med.* **2024**, *97*, 85–92. [CrossRef]
14. Moosavi, S.; Pimentel, M.; Wong, M.; Rezaie, A. Irritable Bowel Syndrome in Pregnancy. *Am. J. Gastroenterol.* **2021**, *116*, 480–490. [CrossRef] [PubMed]

15. Yang, Y.; Wu, N. Gestational diabetes mellitus and preeclampsia: Correlation and influencing factors. *Front. Cardiovasc. Med.* **2022**, *9*, 831297. [CrossRef] [PubMed]
16. Kolodkina, E.V.; Lytaev, S.A.; Galagudza, M.M.; Loginov, A.B.; Pervunina, T.M.; Sitkin, S.I.; Borschchev, Y.Y.; Borschcheva, O.V. Changes in the activity of amylase, alkaline phosphatase and lipase in the placenta and umbilical cord blood in parturient women with gestational diabetes mellitus. *Med. Acad. J.* **2025**, *1*, 62–71.
17. Wu, P.; Ye, Y.; Yang, X.; Huang, Y.; Ye, Y.; Lai, Y.; Ouyang, J.; Wu, L.; Xu, J.; Yuan, J.; et al. Liver biomarkers, lipid metabolites, and risk of gestational diabetes mellitus in a prospective study among Chinese pregnant women. *BMC Med.* **2023**, *21*, 150. [CrossRef]
18. Jarmuzek, P.; Wielgos, M.; Bomba-Opon, D. Placental pathologic changes in gestational diabetes mellitus. *Neuro Endocrinol. Lett.* **2015**, *36*, 101–105. [PubMed]
19. Ivanov, D.O.; Avrel'kina, E.V.; Aleksandrovich, Y.S.; Aleshina, E.I. *Rukovodstvo po Perinatologii*; Inform-Navigator: Saint Petersburg, Russia, 2019.
20. Kothari, S.; Afshar, Y.; Friedman, L.S.; Ahn, J. AGA Clinical Practice Update on Pregnancy-Related Gastrointestinal and Liver Disease: Expert Review. *Gastroenterology* **2024**, *167*, 1033–1045. [CrossRef]
21. Kazma, J.M.; Anker, J.; Allegaert, K.; Dallmann, A. Anatomical and physiological alterations of pregnancy. *J. Pharmacokinet. Pharmacodyn.* **2020**, *4*, 271–285. [CrossRef]
22. King, J.C. Physiology of pregnancy and nutrient metabolism. *Am. J. Clin. Nutr.* **2000**, *71*, 1218–1225. [CrossRef]
23. Godfrey, K.M.; Barker, D.J. Fetal nutrition and adult disease. *Am. J. Clin. Nutr.* **2000**, *71*, 1344–1352. [CrossRef]
24. Morton, A. Investigating gastrointestinal disorders in pregnancy. *Obstet. Med.* **2024**, *17*, 5–12. [CrossRef]
25. Wei, J.; Shen, N.; Shi, C. Exploration of Serum lipid levels during twin pregnancy. *J. Matern. Fetal Neonatal Med.* **2023**, *36*, 2254891. [CrossRef]
26. Penjoyan, G.A.; Model, G.Y.; Korot'ko, G.F. Activity of digestive enzymes in newborns as a prognostic factor for the effectiveness of breastfeeding. *Russ. J. Gastroent. Hepat. Pract.* **2017**, *27*, 39–47.
27. Borovik, T.E.; Moshinsky, I.V.; Mason, V.K. Assessment of intestinal barrier permeability in infants with impaired intestinal function. *Russ. Pediatr. J.* **2012**, *4*, 7–10.
28. Shenderov, B.A. *Medical Microbial Ecology and Functional Nutrition*; Medicine: Moscow, Russia, 2022.
29. Kawabata, A. Modulation by protease—Activated receptors of the rat duodenal motility in virto: Possible mechanisms underlying the evoked contraction and relaxation. *Br. J. Pharmacol.* **1999**, *128*, 865–872. [CrossRef]
30. Radke, R. Similarities and differences in the gastrointestinal tract, gallbladder, and pancreas. *Neurogastroenterology* **2016**, *3*, 93–105.
31. Gorczyca, K.; Obuchowska, A.; Kimber-Trojnar, Z.; Wierzchowska-Opoka, M.; Leszczyńska-Gorzelak, B. Changes in the Gut Microbiome and Pathologies in Pregnancy. *Int. J. Environ. Res. Public. Health* **2022**, *16*, 9961. [CrossRef] [PubMed]
32. Ying, E.; Qian, Y.; Tao, S.; Hang, X.; Xue-Rong, Z.; Hua-Chuan, Z. The relationship between pepsinogen C and gastric carcinogenesis: A transgene and population study. *BMC Cancer* **2023**, *23*, 520. [CrossRef] [PubMed]
33. Riet, H.G.; Van Hockes, J.O. Amylase and lipase values in normal subjects. *Clin. Chim. Acta* **1968**, *19*, 459–467. [CrossRef]
34. Layer, P.; Gloger, G. Fate of pancreatic enzymes in human intestinal lumen in health and pancreatic insufficiency. *Digestion* **1993**, *54*, 4–10. [CrossRef]
35. Salt, W.B.; Schenker, S. Amylase—Its clinical significance: Review of the literature. *Medicine* **1976**, *55*, 269–289. [CrossRef] [PubMed]
36. Kolodkina, E.; Lytaev, S. The Dynamics of Transaminase and Alkaline Phosphatase Activities in the "Mother–Placenta–Fetus" Functional System. *Biomedicines* **2025**, *13*, 626. [CrossRef] [PubMed]
37. Rothman, S.; Liebow, C.; Isenman, L. Conservation of digestive enzymes. *Physiol. Pev.* **2002**, *82*, 1–18. [CrossRef] [PubMed]

biomolecules

MDPI

Article

Is There Need for Pancreatic Enzyme Replacement Therapy in Patients with Exocrine Pancreatic Insufficiency When Using High-Caloric Liquid Diets? Orientating Studies on Praecaecal Digestibility in Pigs with Experimentally Induced Pancreatic Exocrine Insufficiency and Ileocaecal Fistula

Anne Katrin Mößeler [1,2,*], Annette Liesegang [2], Paul Torgerson [3] and Josef Kamphues [1]

1 Institute for Animal Nutrition, University of Veterinary Medicine Hannover, Foundation, D-30173 Hanover, Germany

2 Institute of Animal Nutrition and Dietetics, Vetsuisse Faculty, University of Zurich, CH-8057 Zürich, Switzerland; aliese@nutrivet.uzh.ch

3 Section of Epidemiology, Vetsuisse Faculty, University of Zurich, CH-8057 Zürich, Switzerland

* Correspondence: pancreas@moesseler.de; Tel.: +49-177-63-15-19-4

Abstract: In patients with pancreatic exocrine insufficiency (PEI), focus is primarily placed on fat digestion. Using high-caloric drinks (HCD) is often recommended to avoid malnutrition, but knowledge is limited whether pancreatic enzyme replacement therapy (PERT) is needed. In this study the animal model of pancreatic duct-ligated (PL) and ileocaecal-fistulated minipig was used to determine the praecaecal disappearance rates (pcDR) of the fat and protein of four HCD in controls and PL-pigs with or without PERT. In controls pcDR were high (95.5–96.6% for fat; 70.2–78.6% for protein) while in PL-pigs receiving no PERT the pcDR were significantly lower (fat DR: 47.4–54.3%; protein 22.4–33.5%) despite a high fat pcDR value (84.0%) of one diet. PERT resulted in a normalisation of pcDR of fat and protein with values not differing from controls. This study demonstrates the massive impact of PEI on pcDR, even in HCD typically considered highly digestible. Using PERT is highly recommended in PEI patients using HCD to avoid maldigestion and associated digestive tract symptoms. Optimisation of formulations and galenic preparations of the HCD seems to be necessary as well, as the high fat pcDR of one drink showed that even without PERT high values can be reached.

Keywords: malnutrition; pancreatic exocrine insufficiency; pancreatic enzyme replacement therapy; protein; fat; high-calorific liquid supplements; animal model

1. Introduction

Pancreatic exocrine insufficiency (PEI) is a maldigestion syndrome relevant to both human [1] and veterinary medicine. According to the European guidelines for the diagnosis and treatment of PEI [2], PEI is defined as follows: "A reduction in the exocrine pancreatic secretion and/or intraluminal activity of pancreatic enzymes below the level that allows normal digestion of nutrients. PEI is associated with the malabsorption of nutrients and may result in intestinal symptoms and/or nutritional deficiencies." The aetiology of PEI is diverse. While chronic pancreatitis (CP) and cystic fibrosis (CF) are common causes, acute necrotising pancreatitis, pancreatic cancer, anatomical changes following surgery, and other digestive tract disorders have also been reported as underlying factors. Therefore, pancreatic exocrine insufficiency must be considered a maldigestion syndrome rather than an isolated organ defect [2]. Typical symptoms of PEI are bloating, steatorrhea, and weight loss as well as deficiency in fat-soluble vitamins [1–4], affecting health status but also the quality of life of the patients.

The standard treatment and cornerstone in the treatment of PEI [1–4]—regardless of aetiology—is pancreatic enzyme replacement therapy (PERT), which is performed most frequently using porcine-derived enzymes, although non-porcine alternatives are now available in some countries. Small-sized, enteric-coated pellets are the preferred pancreatin preparations for PEI. The therapeutic goal is the resolution of nutritional deficiencies and the relief of symptoms and signs associated with PEI [2,4]. Enzyme dosage, with or without the addition of a proton pump inhibitor (PPI), should be tailored to the individual [5]. PERT is usually titrated, with the aim of reducing or eliminating steatorrhea [3], but it is worth mentioning that only 71% of international experts prescribes PERT in the case of steatorrhea [6]. PEI ought to be addressed in alignment with recognised clinical guidelines [2,7]. In one study [3], the median PERT dosage used was 96,000 units of lipase with each meal to treat PEI, while a recently published recommendation [4] aims for at least 40,000 USP units lipase per meal in adults and 20,000 USP units lipase per snack.

However, Rijk et al. showed in a recent study [6] based on an international expert survey that despite various published guidelines there is still a considerable lack of consensus and variation in the management of EPI patients with CP. In many malnourished patients with PEI, the use of high-calorific supplemental nutrition is standard practice to ensure adequate nutrient intake. Most human digestibility studies rely on stool samples, which do not distinguish between praecaecal and post-ileal digestive processes. Because post-ileal fat digestion is minimal, stool fat analysis is reliable for assessing fat digestion. However, for protein and starch, post-ileal fermentation can lead to a considerable overestimation of praecaecal digestibility (predominantly enzymatic digestion) and nutritive value [7–13].

Post-ileal crude protein (respectively, N) absorption is of limited benefit—it primarily occurs in the form of ammonia absorption, which is more of a metabolic burden than a nutritional advantage. Therefore, knowledge of praecaecal protein digestion is essential [8,11]. Animal models that allow sampling of digesta at the terminal ileum, in addition to faeces, make it possible to differentiate between enzymatic digestion (praecaecal) and microbial fermentation in the hindgut [8]. In animal nutrition, praecaecal protein digestibility is a global standard for evaluating the nutritional value of protein sources.

Although clinical recommendations for dietary management and PERT in PEI mainly focus on fat digestion and lipase supplementation [2,3], experimental studies in animal models have shown that protein and starch digestibility rates are also impaired [8,10,13,14]. Low praecaecal protein digestibility may contribute to sarcopenia, muscle mass loss, and reduced muscle strength. A recent study [15] demonstrated the importance of adequate PERT in patients with advanced pancreatic cancer to prevent muscle loss. Notably, low-dose PERT (75,000 IU lipase/day) was independently associated with higher odds of muscle loss compared with high-dose therapy—regardless of tumour response, disease stage, or chemotherapy regimen. Patients with PEI may be undertreated [4–6,16]. This highlights the need for sufficient enzyme replacement [4,17,18], with attention not only given to lipase but also to protease supplementation. In the European guidelines for the diagnosis and treatment of pancreatic exocrine insufficiency [2], the occurrence of protein deficiency in case of PEI is also named and high protein food is recommended for EPI patients [4].

Furthermore, a study using the animal model of PEI has demonstrated increased endogenous protein losses [19]. Marked changes in functional proteins may occur in PEI patients. Retinol-binding protein and prealbumin may help to monitor PERT effectiveness in PEI patients [20]. PEI affects these parameters, and maintaining normal levels and functionality of these proteins depends on effective PERT. Currently, published clinical studies evaluating the efficacy and safety of protease enzyme products administered with high-calorie beverages for the prevention of malnutrition are limited.

2. Materials and Methods

2.1. Aim of the Study

This experimental investigation sought to evaluate whether praecaecal digestibility is affected in cases of experimentally induced pancreatic exocrine insufficiency (PEI) when high-calorie liquid supplements are administered. Various diets were used to assess potential differences regarding praecaecal digestibility between products. The actual guidelines for the diagnosis and treatment of pancreatic exocrine insufficiency [2] mention in statement 3.9.2 that "if required, PERT can be added to enteral nutrition; however, its efficacy has not yet been proven". On the other hand, the ESPEN-ESPGHAN-ECFS guidelines on nutrition care for cystic fibrosis [7] state that while a high-calorie, high-fat diet with pancreatic enzyme replacement therapy (PERT) and fat-soluble vitamin supplementation was the standard of nutritional care for CF for decades, no precise information was included regarding high-calorific supplements (drinks). This study also aimed to assess whether PERT is beneficial or even crucial with commercial high-calorific liquid formulas. The focus was placed not only on fat digestion but also on protein digestion, given that protein maldigestion appears to be an underestimated problem in PEI patients [21]. In this orientating study, a high dose PERT was tested as this study aimed to test the principle in general. Therefore, this study did not aim to establish recommendations regarding PERT dosage for PEI patients using HCD.

2.2. Materials and Methods

The studies were conducted using an animal model involving pancreatic duct ligation and ileocaecal fistulation in minipigs. This is a well-established model for investigating the effects of pancreatic exocrine insufficiency (PEI) and pancreatic enzyme replacement therapy (PERT) [10,14,22–26]. The ileocaecal fistula enables the collection of ileal chyme, allowing the quantification of the praecaecal digestibility of nutrients, which is particularly important for assessing starch and protein digestion [10,13].

2.3. Animals

Every effort was undertaken to minimise animal stress and to limit the number of minipigs included in this study. The trial was conducted in accordance with the German animal welfare regulations and with the European Council Directive of 24 November 1986 (86/609/EEC) and was approved by the Ethics Committee of Lower Saxony for the Care and Use of Laboratory Animals (LAVES: Niedersächsisches Landesamt für Verbraucherschutz und Lebensmittelsicherheit; reference: 33.9-42502-04-15/1910). This study used the minimum number of animals required under the 3-R principle and German animal welfare rules.

All animals underwent acclimatisation to handling and sample collection procedures and exhibited no observable indicators of stress or discomfort. Overall, nine adult female Ellegaard® (Dalmose, Denmark) minipigs, fitted with an ileocaecal canula (method described by Tabeling [27]), were used in this study. In five of these animals, PEI was experimentally induced by ligation of the ductus pancreaticus accessorius (PL); the other four with normal exocrine pancreatic function were used as controls (CON). Faecal chymotrypsin concentration was measured in all animals (test kit purchased from Immundiagnostik AG, Wiesenstrasse 4, 64625 Bensheim, Germany, catalogue number K6990) and only minipigs with a chymotrypsin activity < 0.900 U/g faeces were defined and used as

PEI-pigs. In the present study the body weight of the minipigs enrolled was 30.7 ± 4.74 kg (Con) and 27.1 ± 1.21 kg (PL). The animals underwent surgery at least 8 months before the experiments of this study started.

2.4. Liquid Test Diets—Commercial High-Calorific Drinks (HCDs) for Human Consumption

The liquids diets tested in this test were HCDs designed for enteral nutrition of patients with maldigestion and contained at least 1.5 kcal/mL. Each liquid supplement was fed to every animal in a Latin square design, and the animals underwent each trial twice—one time without PERT and one period with a porcine pancreatic enzyme product at a dosage of 300,000 I.U. lipase (accounting for 20,600 I.U. protease per meal). Additionally, all liquid diets were tested in four healthy minipigs with normal exocrine pancreatic function, which were used as controls. The test was conducted as a screening procedure based on the methodology established by Becker [28] and Classen [29], and accordingly, the term "disappearance rate" (DR) was employed in place of "digestibility rate" as referenced by Mößeler et al. 2007 [8]. Phases without PERT for the PL-pigs were reduced to a minimum (this was generally performed 24 h before the screening test to avoid any influence on test results and the screening test itself if PL animals underwent the trial without PERT).

For this study four commonly used liquid nutritional supplements (high-calorific drinks (HCDs)) from different suppliers were selected from a local pharmacy in Germany according to the local selling figures in the year 2016. Of these four HCDs, two (A and B) were rich in fat (9.3 g/100 mL) compared to values of 4.9 and 5.8 g/100 mL in the other diets. Protein content was moderate (5.6 and 6.25 g/100 mL) in two of these HCDs with higher values in the other ones used (9.6 and 10.2 g/100 mL). Among these HCDs, two had an energy density of 1.5 kcal/mL, while the other two (A and B) had an energy density of 2.4 kcal/mL. The nutritional key points of these HCDs are given in Table 1.

Table 1. Nutrient content (g) as well as energy density (per 100 mL) of the commercial high-calorific liquid products designed for enteral nutrition to prevent malnutrition used in this study.

HCD	Fat (Total)	Fat Saturated	Protein	kJ/kcal
A	9.3	0.9	9.6	1010/240
B	9.35	0.83	10.2	1008/240
C	5.8	0.4	5.6	630/150
D	4.9	0.5	6.25	630/150

Each liquid diet was given at an amount of 800–850 mL per meal and nutrient intake was calculated from the analysed values of the diets (including 30 g Methocel™ (Hydroxypropyl-Methylcellulose, Sigma Aldrich Chemie, Taufkirchen, Germany). For two HCDs (A and B), the protein intake was around 81 resp. 85 g while the other HCD resulted in a total intake of around 44 and 50 g (see Table 2).

Table 2. Amount of nutrients per test meal using the commercial high-calorific liquid products designed for enteral nutrition to prevent malnutrition used in this study.

HCD	Dry Matter Intake (g)	Fat Intake (g)	Protein Intake (g)
A	414	75	81
B	422	77	85
C	260	47	44
D	273	42	50

2.5. Test Design

The tests were performed as a screening test established by Becker [28] and Classen [29] to test the efficacy of PERT in the animal model of the ileocaecal-fistulated and pancreatic duct-ligated minipig. The different test diet drinks (including 0.368 g Cr_2O_3 (Sigma Aldrich Chemie, 98% $\leq$ 50 μm) as a marker to calculate disappearance rates) were fed only once (no adaptation period) and chyme collection was performed over 8 h after the first occurrence of the green colour of the chyme indicated the arrival of the test diet at the fistula at the terminal ileum [27]. To ensure a parallel flow of nutrients and marker, 30 g methylcellulose (Methocel™, Sigma Aldrich Chemie, Taufkirchen, Germany) was added to each meal. When animals were included in a trial, the maintenance therapy (regular administration of porcine multienzyme product) for PL-pigs was discontinued the last day prior to the test. At the evening meal before the test, the pigs were fed only 400 mL of a liquid diet (ProvideXtra DRINK™ by Fresenius Kabi Deutschland GmbH, Bad Homburg, Germany) to optimise gastric emptying and to avoid a carryover of nutrients of the diet from the previous evening into the test [28,29]. A period of at least 48 h was performed between two screening tests to ensure a washout of Cr_2O_3 out of the praecaecal gastrointestinal tract. The different liquid test diets were tested in a randomised order with every pig receiving all diets. The control pigs received each test diet once while the PL-pigs underwent the test of each liquid diet twice—one time without PERT and one time with PERT. A coated commercial pancreatic enzyme product of porcine origin was administered at a dosage of 300,000 I.U. lipase (containing 20,600 I.U. protease) per meal. The enzymes were added to the liquid diet. Test diets were taken up rapidly and completely by all pigs.

2.6. Sampling

The ileal chyme was sampled by opening the re-entrant fistula, closing the caecal fistula with a metallic cover containing a membrane. Positioning of flexible bins at the ileal fistula allowed collection of the ileal chyme.

The bins were regularly serviced, with frequent emptying and replacement as necessary. Substitution of water and electrolytes at an amount of 1 d/g collected chyme) was carried out via the membrane of the caecal fistula every two hours according to the method described by Tabeling [29].

2.7. Sample Preparation and Analyses

Sampled ileal chyme was weighed immediately after collection using an analytical scale and then frozen at −20 °C. Afterward, the samples were freeze-dried (freeze dryer: alpha 1–4 LSC, Martin Christ, Gefriertrocknungsanlagen GmbH, Osterode am Harz, Germany), and analysed regarding dry matter, crude fat, and crude protein according to reference [30]; crude fat content was determined by acid hydrolysis and petrol ether extraction using a filter bag technique in an extractor (ANKOM XT15; ANKOM Technology, Macedon, NY, USA) for protein combustion after Dumas. Cr_2O_3 was determined according to the method of Petry and Rapp [31]. The apparent praecaecal disappearance rate (pcD) was calculated according to Becker [27]. For fat as well as protein, the values given represent the apparent digestibility, i.e., the disappearance rate.

2.8. Statistical Analyses

The *t* test was used to compare the praecaecal disappearance rate of fat and protein between controls, PL and PL + PERT groups. Alternatively, where there was evidence of significant departure of the data from a normal distribution, comparisons were made using a Wilcoxen test. All analysis was undertaken in R (R Core team, 2024; R version 4.3.3, R Foundation for Statistical Computing, Vienna, Austria, https://www.R-project.org).

3. Results

3.1. *Effect of PEI and PERT on pcDR of Fat*

The fat pcDR was very high (varying between 95.5% ± 2.20 (HCD B) and 96.6% ± 1.75 (HCD A)) in CON (see Figure 1A,B). PL-pigs showed a significantly lower fat pcDR with values varying between 47.4% ± 14.2 (HCD A) and 54.3% ± 12.6 (HCD B) for three of the HCD tested. There was one exception: the fat pcDR estimated for HCD C was quite high (84.0% ± 9.60) in PL-pigs and the value did not differ significantly from that observed in CON. PERT resulted in a marked increase in fat pcDR with values not significantly differing from CON (being even numerically higher), resulting in values between 96.6 ± 1.17 (HCD D) and 99.0 ± 0.365 (HCD A).

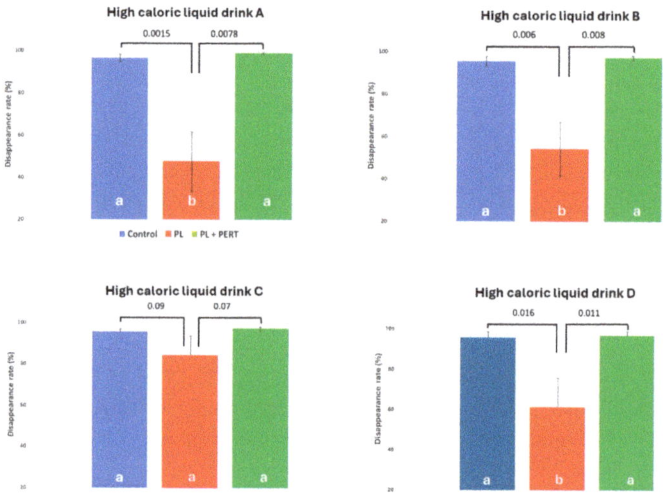

Figure 1. (A–D) Praecaecal disappearance rate of fat in CON (blue bar), PL (red bar) and PL + PERT (green bar) group when different high-calorific liquid diets were fed. Different letters mark significant differences between the treatment groups.

3.2. *Effect of PEI and PERT on Apparent pcDR of Crude Protein*

The apparent pcDR of crude protein was rather high (varying between 70.2% ± 9.19 (HCD D) and 78.6% ± 8.43 (HCD C) in CON (see Figure 2)). For all HCDs, the PEI-pigs showed a significantly lower protein pcDR with values varying between 22.4% ± 3.35 (HCD B) and 33.5% ± 11.6 (HCD C). PERT significantly increased protein pcDR in PL-pigs, with results comparable to CON and even numerically higher for HCD B and D, ranging from 75.3% ± 2.13 (HCD C) to 79.8% ± 2.92 (HCD B).

3.3. *Effect of Liquid Drink on DR of Nutrients*

Crude fat pcDR (%): No significant differences were found for either the controls or the PERT group between the tested HCDs regarding the pcDR of crude fat. (see Table 3) In the PL group, values varied between 47 and 54%, while for one HCD (C), the estimated crude fat pcDR was significantly higher, reaching 84%.

Crude protein DR: No significant differences for apparent pcDR of protein were observed between the four HCD (comparing the values within one treatment group (CON, PL, PL + PERT)). The values did not differ between different diets; for all diets, the values observed in the control and PERT group varied between 70 and 80%, while the protein of all diets was digested to a lower rate of 22 to 32% in the PL group (see Table 4).

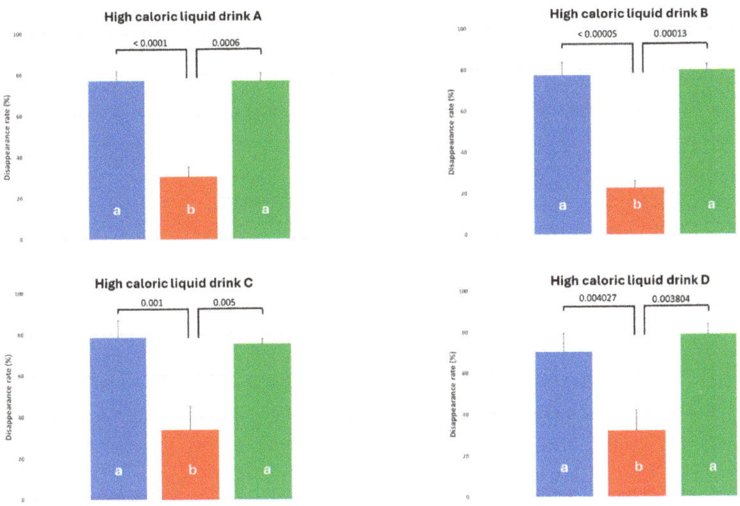

Figure 2. (**A–D**) Apparent praecaecal disappearance rate of protein in controls (blue bar), PL (red bar), and PL + PERT (green bar) group when fed different high-calorific liquid diets. Different letters mark significant differences between the treatment groups.

Table 3. Apparent praecaecal disappearance rate (%) of fat in controls, pancreatic duct-ligated (PL) pigs, and PL-pigs receiving high-dosed pancreatic enzyme replacement therapy.

HCD	CON	PL	PL + PERT
A	96.6 ± 1.75 a	47.4 ± 14.2 a	99.0 ± 0.365 a
B	95.5 ± 2.20 a	54.3 ± 12.6 a	97.2 ± 1.05 a
C	95.7 ± 1.20 a	84.0 ± 9.60 b	97.3 ± 0.820 a
D	96.4 ± 1.47 a	52.8 ± 6.01 a	96.6 ± 1.17 a

Different letters mark significant differences between the tested liquid diets within one group of animals (in one column).

Table 4. Apparent praecaecal disappearance rate (%) of protein in controls, pancreatic duct-ligated (PL) pigs, and PL-pigs receiving high-dosed pancreatic enzyme replacement therapy.

HCD	CON	PL	PL + PERT
A	76.9 ± 4.88	30.2 ± 4.89	76.7 ± 4.26
B	77.1 ± 6.27	22.4 ± 3.53	79.8 ± 2.92
C	78.6 ± 8.43	33.5 ± 11.6	75.3 ± 2.31
D	70.2 ± 9.19	31.9 ± 9.93	78.8 ± 4.88

4. Discussion

4.1. Critical Discussion of the Methods Used

Even though this study was performed in a rather low number of animals, since fistulated animals were used, the results show clear the effects of PEI and PERT on pcDR of fat and protein. The fact that each diet was tested in identical animals and each individual with PEI underwent the tests twice (with or without PERT) minimised individual effects. As expected, the variation was lower in the controls and in the PL animals receiving PERT.

The fixed enzyme dosage used was rather high with 300,000 IU lipase per test meal; as the amount of fat differed between 42 g up to 77 g, the ratio of lipase per g fat varied between 7100 and 3900 IU lipase per g fat. The decision to use a fix dose of enzyme may be questionable—but as a commercial multienzyme product was used, the adaptation of the enzyme dosage according to the fat content of the meal would not result in a comparable protease dosage per g protein either. As this study was an orientating study with a focus on the question of whether enzymes are needed for HCD in patients with PEI and whether the normalisation of digestion is possible in the case of high enzyme dosage, the fixed dosage was chosen.

The amount of HCD per test meal was rather high (800 mL), exceeding typical meal sizes in human patients, resulting in a high nutrient intake, and causing a quite challenging situation regarding digestive processes, as the digestive capacity may have been exhausted. The high volume of the test drink was needed to ensure collection of sufficient amounts of ileal digesta to perform the chemical analyses. The fact that in controls the pcDR was almost complete (above 95% for crude fat for all diets and above 75% for protein in all diets but one) shows that a healthy individual can cope with that amounts of nutrients per meal. As PERT resulted in a normalisation of digestive processes in PL-pigs, with values not differing from controls, it can be concluded that neither enzyme dosage nor absorptive capacity was a limiting factor in this study.

4.2. Clinical Relevance of This Study

An impaired digestion of nutrients is a typical symptom of PEI and is well known for complex meals, but knowledge about HCDs designed to prevent malnutrition is limited, with only few studies available [32]. As expected, the tested HCDs were highly digestible in CON animals with values varying between 95 and 97% for fat and around 70 to 79% for crude protein. On the other hand, the pcDR of fat was, in general, much lower in PL-pigs without PERT but also showed much more variation between the different HCD used. For crude protein the pcDR values were distinctly lower in PL-pigs compared to controls. High-dosed PERT was able to cause a significant improvement and normalisation of pcDR for fat as well as protein, reaching the values of the controls.

As many studies in PEI patients are focussed on fat [32], protein maldigestion resulting in protein deficiency may be an underestimated problem. Protein deficiency may cause muscle loss and impair many functions, as many important molecules in metabolism are proteins [15]. A study recently performed in mice [33] showed that deficiency of dietary protein and amino acids can inhibit the mRNA translation of pancreatic digestive enzymes and therefore PEI and malnutrition can be induced. From the findings of that study and the findings from our own investigations (massively impaired protein digestibility in PEI), it can be assumed that an untreated PEI may occur due to protein deficiency.

4.3. Consequences for Practical Recommendations for Nutritional Support for PEI Patients: Does the Labelled Nutrient Content Reflect the Real Nutrient Uptake?

Furthermore, the present results underline the difficulties of nutritional recommendations for PEI patients, as the nutrient intake may be quite different from the nutrient absorption [9]. To clarify this aspect, a calculation of the amount of digested nutrients and comparison with the intake on 100 mL basis was performed. By considering the pcDR and the nutrient content of the HCD, the nutrient uptake per 100 mL can be calculated (see Table 5). While for CON and PL + PERT the correlation between nutrient intake and digested nutrients was quite good and 95 to 99% of crude fat that was ingested was digested, the ranking of the digested amount of crude fat did not correlate nicely to the crude fat intake in PL. For the two HCDs (A, B) with higher fat content (9.3 g/100 mL), PL-pigs without PERT showed pcDR for fat varying from 47% (HCD A) to 54% (HCD B) while the HCD with lower fat content (HCD C, D) showed comparable (53%; HCD D) or higher values (84%; HCD D). When the total amount of praecaecally-digested crude fat was calculated, the two HCDs (C and D) with 5.8 resp. 4.9 g crude fat per 100 mL had marked differences regarding digested fat (4.87 g/100 mL for HCD C and 2.59 g/100 mL for HCD D).

Table 5. Total amount of crude fat per 100 mL and total amount of crude fat being digested per 100 mL intake (calculated with the mean pc fat DR) in the different groups; in brackets the rank (1 highest, 4 lowest) is given for each column.

HCD	Crude Fat (g/100 mL)	Amount of Digested Crude Fat (g/100 mL)		
		CON	PL	PERT
A	9.30 (2)	8.98 (1)	4.41 (3)	9.20 (1)
C	9.35 (1)	8.93 (2)	5.07 (1)	9.09 (2)
D	5.80 (3)	5.55 (3)	4.87 (2)	5.64 (3)
F	4.90 (4)	4.73 (4)	2.59 (4)	4.73 (4)

It needs to be considered that in PL group the HCD B (with high crude fat content of 9.35 g per 100 mL) and HCD C (5.8 g crude fat content per 100 mL) resulted in a comparable amount of digested fat (5.1 g per 100 mL for HCD B and 4.9 g per 100 mL for HCD C). On the other hand, the high DR of the crude fat in HCD C resulted in a much better utilisation of fat from this diet. If only declared values were considered, HCD C's benefit would be underestimated. Taking this into account, it is obvious that conclusions cannot be drawn directly from the labelled nutrient content to the nutrient intake. The fact that HCD C showed a much better DR leads to the question of whether this can be assumed by the declared composition. As in the other HCDs, oils of plant origin were the basis of this HCD, showing markedly higher pcDR and therefore higher benefit for the patient. As observed in an earlier study using this animal model the fat source, i.e., the fatty acid composition, markedly affects digestibility in PL individuals [11,34]. The melting point of fat had a distinct effect on digestibility of PEI (the lower the melting point, the higher the digestibility rates of the fat in PL-pigs). It can be assumed that beside chemical composition, the preparation or technical aspects are relevant (e.g., emulsification). Considering the labelled ingredients of the tested HCD, the information is limited, as all HCDs contained milk protein and oils of vegetable origin (see Table 6). As all HCDs contained rapeseed and sunflower oil as fat sources (with HCD B and D containing also corn oil), it can be speculated that emulsification is of great interest and may explain the differences in pcDR. Therefore, the fatty acid composition and/or grade of emulsification is assumed to be important for the digestibility of fat and should be in focus when these high-calorific diets are formulated and optimised. For future studies, it seems therefore reasonable to do more precise analyses and characterise the fat and emulsification grade of the HCD.

Table 6. Labelled compounds of the commercial high-caloric liquid products designed for enteral nutrition to prevent malnutrition used in this study.

HCD	Fat Source	Emulsifier	Protein Source	Carbohydrate Source
A	rapeseed oil, sunflower oil	lecithin from soy	cow milk	glucose
B	rapeseed oil, sunflower oil, corn oil	lecithin from soy, E466	cow milk	hydrolysed corn starch, saccharose
C	rapeseed oil, sunflower oil	E471, lecithin from soy	cow milk	maltodextrin, sugar
D	rapeseed oil, corn oil, sunflower oil	E471	cow milk	glucose, saccharose

When looking at the situation regarding the crude protein intake, the differences were smaller (see Table 7). Regarding protein, two of the four HCDs used in this study were higher in protein (9.6 (HCD A) and 10.2 (HCD B) g per 100 mL) than the others (5.6 to 6.3 g). Nonetheless, the HCD with the highest protein content (HCD C, 10.2 g/100 mL) resulted only in a slightly higher (2.29 g vs. 1.88 g/100 mL) amount of digested protein compared to the one (HCD D) with a distinctly lower protein content (5.6 g protein/100 mL). Additionally, even though the crude protein intake with HCD B (10.2 g/100 mL) was markedly higher compared to HCD D (3.55 g per 100 mL drink), the amount of digested crude protein was only 0.28 g/100 mL higher in PL (see Table 7). Therefore, for protein—same as for fat—the labelled nutrient content does not allow to calculate the amount of digested nutrients in the case of untreated PEI: When PERT was used, the labelled fat and protein content and the amount of digested nutrients matched quite well.

Table 7. Total amount of crude protein per 100 mL and total amount of crude fat being digested per 100 mL intake (calculated with the mean pc protein DR) in the different groups; in brackets, the rank (1 highest, 4 lowest) is given for each column.

HCD	Crude Protein (g/100 mL)	Amount of Digested Crude Protein (g/100 mL)		
		CON	PL	PERT
A	9.60 (2)	7.38 (2)	2.90 (1)	7.36 (2)
B	10.2 (1)	7.86 (1)	2.29 (2)	8.14 (1)
C	5.60 (4)	4.40 (4)	1.88 (4)	4.22 (4)
D	6.30 (3)	4.42 (3)	2.01 (3)	4.96 (3)

To sum up, PEI without PERT resulted in a distinct reduced pcDR of fat and protein. The finding of reduced fat digestion is well known and typical for PEI. Furthermore, the use of the animal model of pancreatic duct-ligated and ileocaecal-fistulated pig allows the study of protein digestion in the case of PEI [8,10,13,14,19], where praecaecal digestion was also impaired markedly.

This study clearly underlines the need to use PERT when HCDs are used to prevent malnutrition and to optimise nutritional status although the present data does not allow to give any recommendations regarding enzyme dosage needed. The use of PERT is of special interest as low praecaecal digestibility results in post-ileal fermentation, causing bloating, flatulence, and pain as well as steatorrhea and osmotic diarrhoea. PERT is quite easy to implement when patients are ingesting the HCD orally—but much more challenging when patients receive liquid diets during the night via a percutaneous endoscopic gastrostomy (PEG) tube. The recommendation of the ESPEN-ESPGHAN-ECFS guideline on nutrition care for cystic fibrosis [7] states that "Pancreatic enzymes can be administered orally at the start of enteral nutrition (EN) and during the night if the patient is awake. Enzymes in non-enteric or powder form can be mixed in with the feed if oral intake of enzymes is not possible." This orientating study showed clearly that enzymes are needed for HCD in patients with PEI to normalise digestive processes and should be used whenever possible. These results also show clearly that the benefit of using HCD for the patients depends not only on the labelled nutrient content, but also on the digestibility. As undigested nutrients may cause digestive disorders and impair the wellbeing of the patient, this needs to be considered. An interesting finding was the marked difference between the different HCD tested—while most diets showed comparable results, one HCD showed a much higher pcDR for fat than the others. This result indicates that the HCDs differ regarding the digestibility and that tests for evaluation and optimisation are needed. The animal model used in this study is very valuable to evaluate the effects of PERT and to test different diets; nonetheless, the need for fistulated animals and the quite high level of care limits studies. However, another model that was used to evaluate the protein digestion was described previously [35] and might be an additional interesting method to assess protein digestibility.

5. Conclusions

This experimental study shows the marked effects of PEI on pcDR of fat as well as protein. High-dosed PERT was able to cause a complete normalisation of digestion in individuals with PEI while the HCDs were only digestible to a low extent when no PERT was given (with one exceptional high pcDR of fat for one HCD). The fact that one HCD showed a quite high pcDR of fat when given without PERT raises the question whether HCD can be optimised regarding components and galenic preparation to achieve a high pcDR of fat even without PERT. The pcDR for protein were rather low in PL-pigs receiving no PERT, underlining the need to take protease supplementation into account in the case of PEI.

In conclusion, for the praxis we recommend that different HCDs should be used and the clinical outcome (weight gain, symptoms of the digestive tract due to undigested nutrients entering the hindgut causing bloating or steatorrhea) should be observed carefully. But it must be kept in mind that when high-dosed PERT was used, the values of digestibility were high for all HCDs used in this study and did not differ from those of the controls. It can be assumed that even lower-dosed PERT improves digestibility of nutrients of these HCDs. Therefore, it seems reasonable and necessary to recommend multienzyme PERT whenever HCDs are used in patients with PEI. Although the translation of results from an animal study into clinical practice must always be carried out carefully, this orientating study in the animal model may help to deepen the knowledge and understanding of the need for PERT in PEI patients. This is of special interest as the percentage of international experts that recommend PERT [4] is surprisingly low.

Author Contributions: Conceptualisation, A.K.M. and J.K.; methodology, A.K.M. and J.K.; investigation, A.K.M., data evaluation, A.L. and P.T.; writing—original draft preparation, A.K.M.; writing—review and editing, A.K.M., A.L. and J.K.; visualisation, A.K.M.; supervision, J.K.; project administration, A.K.M.; funding acquisition, A.K.M. and J.K. All authors have read and agreed to the published version of the manuscript.

Funding: This research received no external funding.

Institutional Review Board Statement: Every effort was undertaken to minimise animal stress and to limit the number of minipigs included in this study. The trial was conducted in accordance with the German animal welfare regulations and with the European Council Directive of 24 November 1986 (86/609/EEC) and approved by the Ethics Committee of Lower Saxony for the Care and Use of Laboratory Animals (LAVES: Niedersächsisches Landesamt für Verbraucherschutz und Lebensmittelsicherheit; reference: 33.9-42502-04-15/1910, date of approal: 23 September 2015). This study used the minimum number of animals required under the 3-R principle and German animal welfare rules.

Informed Consent Statement: Not applicable.

Data Availability Statement: The original contributions presented in this study are included in the article/supplementary material. Further inquiries can be directed to the corresponding author(s).

Acknowledgments: We would like to kindly thank all the animal keepers and laboratory staff for supporting this experimental work.

Conflicts of Interest: The authors declare no conflicts of interest. The funders had no role in the design of the study; in the collection, analyses, or interpretation of data; in the writing of the manuscript; or in the decision to publish the results.

Abbreviations

Con	Control
DR	Disappearance rate
HCD	High-calorific drink
pcDR	Praecaecal disappearance rate
PEI	Pancreatic exocrine insufficiency
PERT	Pancreatic enzyme replacement therapy
PL	Pancreatic duct-ligated
PL + PERT	Pancreatic duct-ligated individual receiving pancreatic enzyme replacement therapy

References

1. Domínguez-Muñoz, J.E. Pancreatic exocrine insufficiency: Diagnosis and treatment. *J. Gastroenterol. Hepatol.* **2011**, *26* (Suppl. S2), 12–16. [CrossRef]
2. Dominguez-Muñoz, J.E.; Vujasinovic, M.; de la Iglesia, D.; Cahen, D.; Capurso, G.; Gubergrits, N.; Hegyi, P.; Hungin, P.; Ockenga, J.; Paiella, S.; et al. European guidelines for the diagnosis and treatment of pancreatic exocrine insufficiency: UEG, EPC, EDS, ESPEN, ESPGHAN, ESDO, and ESPCG evidence-based recommendations. *United Eur. Gastroenterol. J.* **2025**, *13*, 125–172. [CrossRef]
3. Min, M.; Patel, B.D.; Han, S.; Bocelli, L.D.; Kheder, J.; Vaze, A.; Wassef, W. Exocrine Pancreatic Insufficiency and Malnutrition in Chronic Pancreatitis: Identification, Treatment, and Consequences. *Pancreas* **2018**, *47*, 1015–1018. [CrossRef] [PubMed]
4. Whitcomb, D.C.; Buchner, A.M.; Forsmark, C.E. AGA Clinical Practice Update on the Epidemiology, Evaluation, and Management of Exocrine Pancreatic Insufficiency: Expert Review. *Gastroenterology* **2023**, *165*, 1292–1301. [CrossRef] [PubMed]
5. Lewis, D.M.; Rieke, J.G.; Almusaylim, K.; Kanchibhatla, A.; Blanchette, J.E.; Lewis, C. Exocrine Pancreatic Insufficiency Dosing Guidelines for Pancreatic Enzyme Replacement Therapy Vary Widely Across Disease Types. *Dig. Dis. Sci.* **2024**, *69*, 615–633. [CrossRef]
6. de Rijk, F.E.M.; van Veldhuisen, C.L.; Besselink, M.G.; van Hooft, J.E.; van Santvoort, H.C.; van Geenen, E.J.; Hegyi, P.; Löhr, J.-M.; Dominguez-Munoz, J.E.; de Jonge, P.J.F.; et al. Diagnosis and treatment of exocrine pancreatic insufficiency in chronic pancreatitis: An international expert survey and case vignette study. *Pancreatology* **2022**, *22*, 457–465. [CrossRef]
7. Wilschanski, M.; Munck, A.; Carrion, E.; Cipolli, M.; Collins, S.; Colombo, C.; Declercq, D.; Hatziagorou, E.; Hulst, J.; Kalnins, D.; et al. ESPEN-ESPGHAN-ECFS guideline on nutrition care for cystic fibrosis. *Clin. Nutr.* **2024**, *43*, 413–445. [CrossRef]
8. Möβeler, A.; Tabeling, R.; Gregory, P.C.; Kamphues, J. Compensatory digestion of fat, protein and starch (rates and amounts) in the large intestine of minipigs in case of reduced precaecal digestion due to pancreatic duct ligation—A short review. *Livest. Sci.* **2007**, *109*, 50–52. [CrossRef]
9. Mößeler, A.K. Variety of effects of exocrine pancreatic insufficiency—more than just steatorrhea! In *New Findings from Studies Using the Animal Model of the Pancreatic Duct Ligated Pig and Subsequent Recommendations for Dietetic Measures in Patients*; English Summary, Habilitation in German; Hanover Library of the University of Veterinary Medicine: Hannover, Germany, 2016. Available online: https://nbn-resolving.org/urn:nbn:de:gbv:95-h2785 (accessed on 3 June 2025).
10. Mößeler, A.; Kamphues, J. Black-Box Gastrointestinal Tract—Needs and Prospects of Gaining Insights of Fate of Fat, Protein, and Starch in Case of Exocrine Pancreatic Insufficiency by Using Fistulated Pigs. *Nutrients* **2017**, *9*, 150. [CrossRef]
11. Kramer, Nils. Investigations on the Precaecal Digestibility of Starch and Fat from Different Origin in Pancreatic Duct Ligated, Ileocaecally Fistulated Minipigs. Ph.D. Thesis, University of Veterinary Medicine Hanover, Hannover, Germany, 2010; Abstract auf Englisch bei. Available online: https://elib.tiho-hannover.de/servlets/MCRFileNodeServlet/etd_derivate_00001238/kramern_ss10.pdf (accessed on 3 June 2025).
12. Schubert, D.C.; Mößeler, A.; Ahlfänger, B.; Langeheine, M.; Brehm, R.; Visscher, C.; El-Wahab, A.A.; Kamphues, J. Influences of exocrine pancreatic insufficiency on nutrient digestibility, growth parameters as well as anatomical and histological morphology of the intestine in a juvenile pig model. *Front. Med.* **2022**, *9*, 973589. [CrossRef]
13. Redeker, C.; MoBeler, A.; Kamphues, J. Effects of high/low fat meals on prececal digestibility of protein and starch using recommended enzyme dosages (based on lipase and fat) of a multienzyme product—The pancreas duct ligated minipig as a model for exocrine pancreatic insufficiency. In Proceedings of the 18th Congress of the European Society of Veterinary and Comparative Nutrition (ESVCN), Utrecht, The Netherlands, 11–13 September 2014.
14. Tabeling, R.; Gregory, P.; Kamphues, J. Studies on nutrient digestibilities (pre-caecal and total) in pancreatic duct-ligated pigs and the effects of enzyme substitution. *J. Anim. Physiol. Anim. Nutr.* **1999**, *82*, 251–263.
15. Klassen, P.N.; Mazurak, V.C.; Baracos, V.; Martin, L.; Ghosh, S.; Kasnik, J.; Sawyer, M.B. Dose optimization of pancreatic enzyme replacement therapy is essential to mitigate muscle loss in patients with advanced pancreatic cancer and exocrine pancreatic insufficiency. *Clin. Nutr.* **2024**, *43*, 1900–1906. [CrossRef]

16. Sikkens, E.C.M.; Cahen, D.L.; van Eijck, C.; Kuipers, E.J.; Bruno, M.J. Patients with exocrine insufficiency due to chronic pancreatitis are undertreated: A Dutch national survey. *Pancreatology* **2012**, *12*, 71–73. [CrossRef]
17. Berry, A.J. Pancreatic enzyme replacement therapy during pancreatic insufficiency. *Nutr. Clin. Pract.* **2014**, *29*, 312–321. [CrossRef] [PubMed]
18. Phillips, M.E.; Hopper, A.D.; Leeds, J.S.; Roberts, K.J.; McGeeney, L.; Duggan, S.N.; Kumar, R. Consensus for the management of pancreatic exocrine insufficiency: UK practical guidelines. *BMJ Open Gastroenterol.* **2021**, *8*, e000643. [CrossRef] [PubMed]
19. Mösseler, A.; Loock, H.; Classen, J.; Gregory, P.; Kamphues, J. Endogenous nitrogen losses (praecaecal and total) in case of exocrine pancreatic insufficiency (induced by pancreatic duct ligation) in pigs—Used as a model for humans. *Pancreatology* **2013**, *13*, S24.
20. Lindkvist, B.; Phillips, M.E.; Domínguez-Muñoz, J.E. Clinical, anthropometric and laboratory nutritional markers of pancreatic exocrine insufficiency: Prevalence and diagnostic use. *Pancreatology* **2015**, *15*, 589–597. [CrossRef]
21. Airinei, G.; Gaudichon, C.; Bos, C.; Bon, C.; Kapel, N.; Bejou, B.; Raynaud, J.J.; Luengo, C.; Aparicio, T.; Levy, P.; et al. Postprandial protein metabolism but not a fecal test reveals protein malabsorption in patients with pancreatic exocrine insufficiency. *Clin. Nutr.* **2011**, *30*, 831–837. [CrossRef]
22. Imondi, A.R.; Stradley, R.P.; Wolgemuth, R. Enzyme Replacement Therapy in the Pancreatic Duct Ligated Swine. *Proc. Soc. Exp. Biol. Med.* **1972**, *141*, 367–372. [CrossRef]
23. Abello, J.; Pascaud, X.; Simoes-Nunes, C.; Cuber, J.C.; Junien, J.L.; Rozé, C. Total pancreatic insufficiency in pigs: A model to study intestinal enzymes and plasma levels of digestive hormones after pancreatic supplementation by a whole pancreas preparation. *Pancreas* **1989**, *4*, 556–564.
24. Gregory, P.C.; Tabeling, R.; Kamphues, J. Growth and digestion in pancreatic duct ligated pigs. Effect of enzyme supplementation. In *Biology of the Pancreas in Growing Animals*; Pierzynowski, S.G., Zabielski, R., Eds.; Elsevier Science Health Science: St Louis, MO, USA, 1999; pp. 381–394.
25. Ishihara, R.; Gregory, P.; Koch, H.F.; Kolleck, S.; Mößeler, A.; Kamphues, J.; Uchida, I. Comparison Study of Pharmacological Efficacy of Pancreatic Digestive Enzyme Replacement (LipaCreon Granules) and of Digestive Enzymes (Berizym Combination Granules and Excelase Combination Granules) on Digestibility of Fat, Protein and Starch in a Minipig Model of Pancreatic Exocrine Insufficiency (in vivo). *J. New Rem. Clin.* **2012**, *61*, 1044–1053.
26. Gregory, P.C.; Hoffmann, K.; Kamphues, J.; Möeler, A. The pancreatic duct ligated (mini)pig as model for pancreatic exocrine insufficiency in man. *Pancreas* **2016**, *45*, 1213–1226. [CrossRef]
27. Tabeling, R. Untersuchungen am Pankreasgangligierten Schwein zu Effekten Einer Enzympräparation auf die Nährstoffverdaulichkeit (praecaecal/in toto). Ph.D. Thesis, University of Veterinary Medicine Hanover, Hanover, Germany, 1998.
28. Becker, C. Development of A Screening-Test to Assess Activity of Substituted Proteolytic and Amylolytic Enzymes by Using Pancreatic Duct Ligated Minipigs. Ph.D. Thesis, University of Veterinary Medicine Hanover, Hannover, Germany, 2005. Available online: http://elib.tiho-hannover.de/dissertations/beckerc_ws05.html (accessed on 3 June 2025).
29. Classen, J. Development of A Novel Screening Test to Assess the Efficacy of Multi-Enzyme Products by Using pancreatic Duct Ligated, Ileo-Caecal Fistulated Minipigs. Doctoral Dissertation, University of Veterinary Medicine Hanover, Hannover, Germany, 2008. Available online: https://elib.tiho-hannover.de/servlets/MCRFileNodeServlet/etd_derivate_00001684/classenj_ss08.pdf (accessed on 3 June 2025).
30. Naumann, C.; Bassler, R. Methoden der landwirtschaftlichen Forschungs- und Untersuchungsanstalt, Biochemische Untersuchung von Futtermitteln. In *Methodenbuch III*; VDLUFA: Darmstadt, Germany, 2012.
31. Petry, H.; Rapp, W. Zur Problematik der Chromoxidbestimmung in Verdauungsversuchen. *Z. Tierphysiol. Tierernähr. Futtermittelkunde* **1970**, *27*, 181–189. [CrossRef]
32. Hauenschild, A.; Ewald, N.; Klauke, T.; Liebchen, A.; Bretzel, R.G.; Kloer, H.; Hardt, P.D. Effect of liquid pancreatic enzymes on the assimilation of fat in different liquid formula diets. *J. Parenter. Enteral Nutr.* **2008**, *32*, 98–100. [CrossRef]
33. Sans, M.D.; Crozier, S.J.; Vogel, N.L.; D'Alecy, L.G.; Williams, J.A. Dietary Protein and Amino Acid Deficiency Inhibit Pancreatic Digestive Enzyme mRNA Translation by Multiple Mechanisms. *Cell. Mol. Gastroenterol. Hepatol.* **2021**, *11*, 99–115. [CrossRef]
34. Mößeler, A.; Kramer, N.; Loock, H.; Kalla, K.; Classen, J.; Gregory, P.C.; Kamphues, J. New findings regarding praecaecal fat digestion in pancreatic duct ligated pigs. In Proceedings of the 13th Congress of the European Society of Veterinary and Comparative Nutrition (ESVCN), Zürich, Switzerland, 6–8 September 2010; p. 45.
35. Pierzynowski, S.G.; Wychowański, P.; Szczesny, W.; Galloto, R.; Zaworski, K.; Szkopek, D.; Woliński, J.; Donaldson, J.; Pierzynowska, K. Algorithm and ninhydrin method allow for measurement of the postprandial appearance of peptides in blood. *Sci. Rep.* **2025**, *15*, 19704. [CrossRef]

biomolecules

Article

Dependency of Glucose Homeostasis on Pancreatic Enzymes with Special Reference to Amylase; Study on Healthy and Exocrine Pancreatic Insufficient Pigs

Piotr Wychowański [1,2,3,*], Stefan G. Pierzynowski [4,5,6], Kamil Zaworski [7], Robert Gallotto [8], Dominika Szkopek [9,10], Jarosław Woliński [9], Janine Donaldson [5,11], Tomasz Jacek [10] and Kateryna Pierzynowska [4,5,6,7,10,*]

1 Department of Head and Neck and Sensory Organs, Division of Oral Surgery and Implantology, Institute of Clinical Dentistry, Gemelli Foundation for the University Policlinic, Catholic University of the "Sacred Heart", 00168 Rome, Italy
2 Department of Interventional Dentistry, Collegium Medicum, Nicolaus Copernicus University, 87-100 Bydgoszcz, Poland
3 Department of Periodontology and Oral Diseases, Medical University of Warsaw, 02-091 Warsaw, Poland
4 Department of Biology, Lund University, 22362 Lund, Sweden; stefan.pierzynowski@biol.lu.se
5 Anara AB, 23132 Trelleborg, Sweden
6 Department of Medical Biology, Institute of Rural Health, 20-090 Lublin, Poland
7 Department of Animal Physiology, The Kielanowski Institute of Animal Physiology and Nutrition, Polish Academy of Sciences, 05-110 Jabłonna, Poland; k.zaworski@ifzz.pl
8 Anagram Therapeutics, Inc., Natic, MA 01701, USA; rgallotto@anagramtx.com
9 Large Animal Models Laboratory, The Kielanowski Institute of Animal Physiology and Nutrition, Polish Academy of Sciences, 05-110 Jabłonna, Poland; d.szkopek@ifzz.pl (D.S.); j.wolinski@ifzz.pl (J.W.)
10 Research Institute of Animal Husbandry, 111121 Balice, Poland; tomasz.jacek@iz.edu.pl
11 Department of Physiology, School of Biomedical Sciences, Faculty of Health Sciences, University of the Witwatersrand, Johannesburg 2193, South Africa; janine.donaldson@wits.ac.za
* Correspondence: piotrwychowanski@wychowanski.pl (P.W.); kateryna.pierzynowska@biol.lu.se (K.P.)

Abstract

We aimed to highlight the roles of the pancreatic enzymes, with special reference to amylase, on glucose homeostasis in healthy pigs and in pigs with exocrine pancreatic insufficiency (EPI). Healthy pigs fed a high-fat diet (HFD) were subjected to mixed meal tolerance tests (MMTTs) and pancreatic enzyme treatments, and then blood glucose and insulin concentrations were determined. Following the development of surgically induced EPI, the same experiment was then repeated on the pigs. A significantly lower net postprandial glycemic response was observed in pigs with EPI compared to healthy pigs. Net postprandial glycemic response was not affected by enzyme supplementation during the MMTTs in healthy pigs, but it was affected by adaptation to macronutrient components of the MMTT test meal, both in healthy and EPI pigs. Net postprandial glycemic response and insulin release curves reached higher levels in Creon-treated EPI pigs compared to amylase-treated EPI pigs. In summary, glucose homeostasis mechanisms in EPI pigs were downregulated compared to healthy animals. Creon supplementation during EPI significantly increased postprandial glucose level, while amylase treatment had the opposite effect, which could be explained by its metabolic actions.

Keywords: mixed meal tolerance test; MMTT; glucose metabolism; insulin release; amylase; pancreatic enzyme replacement therapy; PERT

https://doi.org/10.3390/biom16010172

1. Introduction

In clinical practice, different oral and parenteral nutritional strategies are currently used to estimate glucose homeostasis to introduce the correct medication or preventive treatment, as well as adequate nutritional support, as needed [1–3]. Glucose homeostasis is the continuous process that stabilizes fasting blood glucose to physiological levels between 80–100 mg/dL in most mammals, including humans, in response to changes in internal (expenditure) and external (absorption) conditions. Insulin and glucagon are commonly recognized as key regulators of glucose homeostasis [4,5].

Other factors that influence glucose homeostasis, such as the macronutrient content, kind of diet, and the adaptation to dietary macronutrients per se [6], are less well recognized. Digestive pancreatic enzymes with possible regulatory features are sometimes recognized as secondary players in glucose homeostasis, but are often neglected. However, it is generally accepted that specific diets constructed for patients are effective in promoting health [7–9]. Over the last few decades, there has been an explosion of dietary guidelines, recommendations, and formulas that are supposed to alleviate the main symptom (hyperglycaemia) of different congenital, metabolic, and/or age-related sicknesses, e.g., cystic fibrosis (CF), type 2 diabetes mellitus (DT2), metabolic syndrome, obesity, etc.

It is important to recognize that the nutritional effectiveness of various diets is highly dependent on the digestion and absorption of the macronutrient components. The qualitative and quantitative composition of pancreatic enzymes, as well as other enzymes of the gastrointestinal tract, including brush border enzymes, plays a primary role in the digestion of dietary macronutrients [10–12].

Pigs represent the accepted animal model for the study of human diseases and have been used for decades, e.g., due to their similarity to humans with regard to their immunology, anatomical size and structure, as well as their physiology [13,14]. The similarities between humans and pigs, specifically with regard to their gastrointestinal structure and function, resulting in comparable digesta transit times, nutrient absorptive processes, gut microbiota, etc., make the pig a more adequate, relevant animal model for humans, compared to the rat [14,15]. Recent studies performed on pigs highlighted the regulatory role of amylase in metabolism and its modulatory features, which are crucial for glucose homeostasis [16]. Amylase, administered orally, was shown to exhibit anti-incretin activity by reducing insulin secretion during an oral glucose tolerance test (OGTT) [16]. The results suggest that oral microbial amylase improves glucose homeostasis by reducing insulin resistance, and it (amylase) exhibits insulin-like features [16].

The benefit or rationality of amylase supplementation in people with exocrine pancreatic insufficiency (EPI) is often overlooked. Pancreatic amylase maintains physiological glucose utilization and, at the same time, reduces insulin release, which can protect pancreatic beta cells from exhaustion during long-lasting hyperinsulinemia [17]. Additionally, microbial amylase used in previous experiments reduced insulin resistance, thus protecting the host against the development of diabetes [18].

The main aim of the current study was to elucidate the specificity of amylase supplementation on glucose homeostasis as compared to Creon, a porcine-derived enzyme mixture commonly used in the treatment of EPI. We also attempted to emphasize the influence of test meal nutritional components on postprandial glycaemia and the importance of the use of the mixed meal tolerance test (MMTT) for the assessment of postprandial glycemic response, while simultaneously highlighting the modulatory action of pancreatic enzymes on the net postprandial glycemic response.

https://doi.org/10.3390/biom16010172

2. Materials and Methods

2.1. Animals

The present study was performed in strict accordance with the recommendations in the Guide for the Care and Use of Laboratory Animals of the National Institutes of Health and in accordance with ARRIVE guidelines. All efforts were made to minimize animal suffering. The study was approved by the Second Local Ethics Committee for Animal Experimentation in Warsaw, Poland (approval no. WAW2/025/2022, approval date 16 February 2022). The experiment was performed on 18 crossbred ((Polish Landrace × Yorkshire) × Hampshire)) pigs (*Sus scrofa domesticus*), purchased from a local herd (Karniewek, Poland), a mix of both genders (50%:50%, males were castrated on day 3 postpartum), aged eight weeks, and with a body weight of 15 ± 2.3 kg at the beginning of the study. The animals were randomly divided into 3 groups (Creon, amylase, and control groups); $n = 6$ pigs in each group. The experiments were performed on the pigs with their pancreas intact and then repeated on the same pigs after pancreatic duct ligation surgery and the development of EPI. The pigs were maintained on a 12 h day–12 h night cycle, with lights on from 06.00 to 18.00 h. The pigs were individually housed in pens (following surgery) for the whole duration of the experimental period. The individual pens were equipped with a feeding trough, drinking nipple, and constant heating lamp (150 W). The pigs were allowed to move freely within their pens and had visual and olfactory contact with each other.

Sample size was estimated using G*Power software [19], version 3.1.9.4, for a one-way ANOVA at $\alpha = 0.05$ with 95% power, assuming f (effect size) = 1.2, for three study groups. The calculation yielded the total number of animals as fourteen. However, we have increased the total number of animals up to 18 to keep the number sufficient for reaching the desired statistical power even in the case of animal loss.

2.2. Feed

During the study, pigs were adapted to a high-fat diet (HFD) (Kcynia, Morawski Plant, Poland, containing approximately 17.5% crude protein, 3.5% crude fiber, 20% crude fat, 50% starch, 4% ash and 5% water; the detailed composition is provided in Table 1) in an amount equivalent to 4% of their body weight daily, with 1% given at the morning meal (09:00–10:00 hr) and 3% at the afternoon meal (17:00–18:00). The HFD is often used to mimic the human diet, in terms of fat content, in metabolic studies using porcine models [20,21]. The feed was developed to fulfil the nutritional requirements of piglets with a body weight between 11 and 25 kg, according to the National Research Council requirements [22], with considerations for the experimental conditions and pig needs. Upon arrival at the experimental unit, the pigs were fed a cereal-based, pelleted, standard pig diet (Kcynia, Morawski Plant, Poland) and abruptly changed to an HFD on the first day of adaptation (Figure 1).

Table 1. Composition of the high-fat diet (HFD) (Kcynia, Morawski Plant, Poland).

Energy and Chemical Composition (per 1 kg Feed)	
Energy, MJ	16.8 (4012.7 kcal)
Water, g	50.0
Crude protein, g	175.0
Carbohydrates, g	500.0
Crude fat, g	200.0
Ashes, g	40.0
Crude fiber, g	35.0

https://doi.org/10.3390/biom16010172

Table 1. *Cont.*

Energy and Chemical Composition (per 1 kg Feed)	
Amino Acids (per 1 kg feed)	
Cysteine + Methionine, g	8.1
Threonine, g	8.7
Lysine, g	13.9
Leucine, g	14.2
Methionine, g	4.7
Total Nitrogen, g	38.8
Minerals (per 1 kg feed)	
Calcium, g	11.0
Chloride, g	3.2
Phosphorus, g	9.0
Potassium, g	3.1
Sodium, g	2.9
Magnesium, g	0.4
Iodine, mg	0.15
Iron, mg	100.0
Manganese, mg	3.5
Zinc, mg	85.0
Copper, mg	4.9
Selenium, mg	0.23
Vitamins (per 1 kg feed)	
Vitamin A, IU	2250
Vitamin D, IU	375
Vitamin E, IU	11
Vitamin K, mg	1.1
Biotin, mg	0.05
Niacin, mg	30.0
Pantothenic acid, mg	10.0
Riboflavin, mg	3.5
Thiamin, mg	1.0
Vitamin B_6, mg	3.0
Vitamin B_{12}, µg	15.5

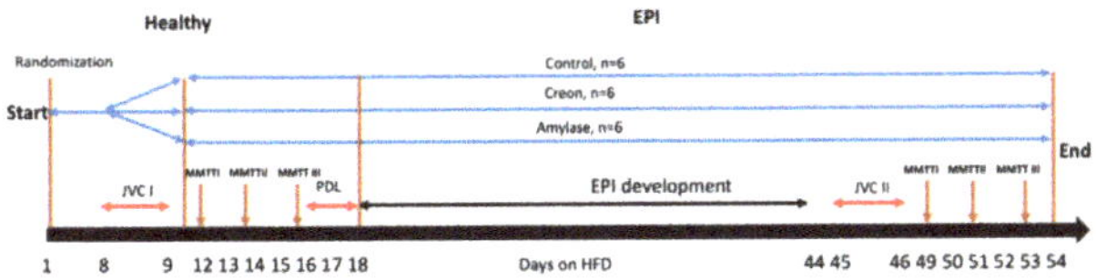

Figure 1. Experimental design and treatments. EPI—exocrine pancreatic insufficiency; JVC—jugular vein catheter implantation; MMTT—mixed meal tolerance test; PDL—pancreatic duct ligation.

2.3. Selected Substrates for MMTT

The following substances were chosen to create the acute test meal used for the MMTTs to mimic a standard high-fat meal (Table 2): docosahexaenoic acid (DHA) and eicosapentaenoic acid (EPA)-Kinoko Life 2000 mg Omega-3 softgels; 500 mg EPA and 250 mg DHA per capsule (Kinoko Life, Malaga, Spain), Whey-California Gold Nutrition 100% Whey Protein Isolate, Unflavored (California Gold Nutrition, Irvine, CA, USA), Potato starch-Roots Circle Gluten-Free Potato Starch (Verno Goods LLC, Elizabeth, NJ, USA).

https://doi.org/10.3390/biom16010172

Table 2. Composition of mixed meal tolerance tests (MMTTs) offered as morning meals, which accounted for 1% of the daily requirements of pigs weighing approximately 16 kg. The meals were isoenergetic. In between experiments, pigs were fed the HFD, 1% and 3%, respectively, for morning and evening meals.

	MMTT I **HFD**	**MMTT II** **HFD + SUB**	**MMTT II** **SUB**
High Fat Diet Components	28 g cereal protein+ 32 g fat–rapeseed oil+ 79 g starch cereals+ 6 g crude fiber+ 8 g ash+ 7 g water= 160 g	18 g cereal protein+ 20 g rapeseed oil+ 50 g starch cereals+ 3 g crude fiber+ 4 g ash+ 5 g water= 100 g	0
Substrate (SUB) Components	0	4 g DHA/EPA+ 10 g whey+ 32 g potato starch+ 8 g soya oil 3 g ash+ 3 g water= 60 g	4 g DHA/EPA+ 28 g whey+ 85 g potato starch+ 28 g soya oil 8 g ash+ 7 g water= 160 g

2.4. Enzymes

A microbial-derived alpha-amylase (Aspergillus oryzae, 4000 Units/dose, Amano Enzymes, Nishiki, Naka-Ku, Nagoya City, Aichi Prefecture, Japan) or porcine-derived enzyme mixture of Creon® 25,000 (Abbot, Chicago, IL, USA; lipase (25,000 PhEur units), protease (1000 PhEur units) and amylase (18,000 PhEur units)) was administered to the pigs with the morning and evening meals, depending on the experimental design randomization. Pigs received a total of 100,000 lipase units/Creon® dose during the study. Like pancreatic α-amylase, the microbial amylase studied hydrolyzes α-1,4-glucosidic linkages as the initial step in the hydrolysis of dietary starch, glycogen, and related polysaccharides. The control group (n = 6) was fed HFD alone; the Creon group (n = 6) was fed HFD + Creon® (2 $\times$ 100,000 units daily); the amylase group was fed HFD + amylase (2 $\times$ 4000 units daily) at the morning and evening meals, respectively, for the entire duration of the experiment. A detailed study design is shown in Figure 1.

2.5. Blood Sampling

Blood samples were collected during the MMTT via a jugular vein catheter, at one hour and then again at one minute prior to feeding, and then at 5, 15, 30, 45, 60, and 120 min after feeding and transferred to BD Vacutainer® glass Aprotinine K3EDTA tubes (BD Diagnostics, Franklin Lakes, NJ, USA). The blood samples were immediately placed on ice before they were centrifuged at 3000$\times$ g for 15 min at 4 °C, and plasma was separated and stored at −80 °C until further analysis.

Blood glucose concentrations were measured directly following blood sampling using a glucometer and test strips (Accu-Chek Aviva, Roche Diagnostics, Mannheim, Germany). Plasma insulin concentrations were measured using a porcine insulin ELISA kit (cat# 10-1200-01; Mercodia, Uppsala, Sweden).

2.6. Experimental Flow

From the first day, all pigs were randomised to either the control, Creon or amylase groups and adapted to the HFD. Additionally, animals from the Creon and amylase groups were adapted to their respective enzyme supplementation during morning and evening

https://doi.org/10.3390/biom16010172

meals (Figure 1). All pigs underwent jugular vein catheter implantation (JVCI) [23] on days 8 and 9. Starting from day 12, and every second day thereafter, the respective MMTTs were performed on healthy pigs (Figure 1). The MMTTs were performed following an overnight fast and took place at the same time and in the same manner as the pigs' normal morning meal. On days 17 and 18, all pigs were subjected to pancreatic duct ligation (PDL) surgery, as previously described [24]. After EPI development (defined as growth retardation, hyperphagia, and lowered immunoreactive trypsin in the blood, as measured 3–4 weeks after surgery), on days 45 and 46, complementary jugular vein catheters (JVC II) were implanted.

Starting from day 49, and every second day thereafter, identical MMTTs, with identical treatment and experimental order as those performed on the pigs in a healthy state, were then performed on the pigs in the EPI state.

2.7. Statistical Analysis

Statistical analysis was performed on the data generated from this study using an ordinary one-way ANOVA for normally distributed datasets or a Kruskal–Wallis test when data was not normally distributed. To report differences in medians between groups, a Kruskal–Wallis test was performed, followed by a pairwise Mann–Whitney U test. The actual differences between medians, as well as the Hodges–Lehmann difference is reported. The data distribution was assessed using the Shapiro–Wilk normality test. Outliers within data sets were identified using the ROUT method of regression (Q = 0.05%). A mixed effects analysis was conducted to examine the effects of the MMTT meal composition and enzymatic treatment on net postprandial glycemic response. All the analyses were carried out using GraphPad Prism 10.0, San Diego, USA. Data was not corrected for multiple comparisons as the study design included planning for the measurable outcomes [25]. Differences were considered significant if $p \leq 0.05$; differences were considered as a trend when $p \leq 0.1$ ($0.05 \leq p \leq 0.10$); data with a Gaussian distribution are expressed as mean ± standard deviation (SD); data with a non-Gaussian distribution are expressed as median ± interquartile range (IQR). Area under the curve (AUC) data is baseline adjusted.

3. Results

The median values of baseline-adjusted glucose AUCs describe the net absorption with basal metabolic values subtracted. AUC values are always directly proportional to absorption capacity and inversely proportional to expenditure/metabolic capacity during test time. AUCs obtained during different MMTTs are presented in Figure 2. Glucose AUC values obtained from healthy pigs were significantly higher when compared to those seen in the same pigs after EPI development. The median for postprandial glycemic AUCs was approximately 384 in EPI pigs and 900 in healthy pigs, with an actual median difference of −516.0, and a Hodges–Lehmann difference of −342.9 (p = 0.0140), independently of enzyme treatment or test meal composition.

The glucose AUCs in healthy pigs were not affected by Creon or microbial amylase supplementation (Figure 3A), but they were significantly affected by the type of macronutrients used during the various MMTTs (Figure 3C). The median postprandial glucose response in healthy pigs, during the MMTT with components to which the pigs were not adapted to at all (substrate (SUB)), was 427.5, while the median observed during the MMTT with the HFD (100% adaptation), was 1335.0, with an actual median difference of −907.5 and a Hodges–Lehmann difference of −967.5). The median observed during the MMTT with the HFD + SUB (60% adaptation) was 1089.0, with an actual median difference of −661.5 and a Hodges–Lehmann difference of −672.5 (Figure 3C).

https://doi.org/10.3390/biom16010172

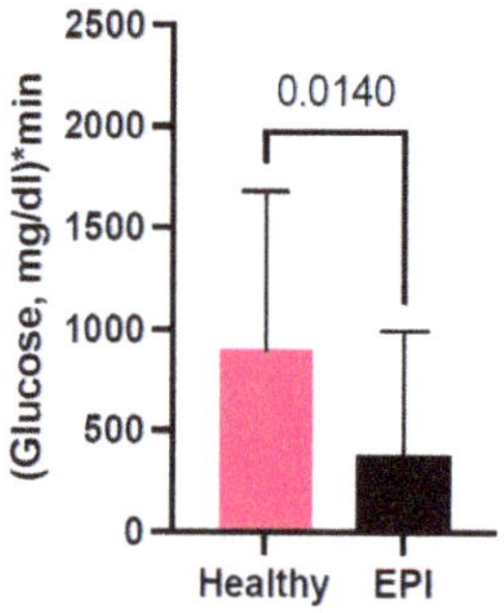

Figure 2. Net postprandial glycemic response AUCs during all MMTTs, with and without enzyme supplementation, 50 values per group. Area under the curve (AUC) data is baseline adjusted. Data are expressed as median ± interquartile range (IQR). Differences between the results were considered significant at $p < 0.05$. p-values are given with the result bars.

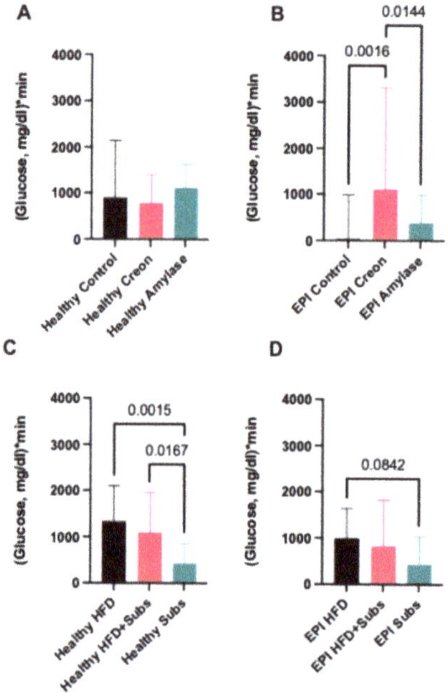

Figure 3. (**A–D**) Combined net postprandial glycemic response AUC values (14 values per result bar) obtained during MMTT in healthy and EPI pigs. (**A**) Summary of glucose AUCs describing the effects of enzymatic treatments in healthy animals on net postprandial glycemic response during all MMTTs; (**B**) summary of glucose AUCs describing the effects of enzymatic treatments in EPI animals on net postprandial glycemic response during all MMTTs; (**C**) summary of glucose AUCs describing the effects of different MMTTs composition in healthy animals on net postprandial glycemic response; (**D**) summary of glucose AUCs describing the effects of different MMTT compositions in EPI animals on net postprandial glycemic response. Data are expressed as median ± interquartile range (IQR). Differences between the results were considered significant at $p < 0.05$; $p < 0.1$ was considered a trend; p-values are given with the result bars.

In the same animals, after the development of EPI, during identical MMTTs, Creon treatment significantly increased the glucose AUC compared to that observed in control EPI pigs (actual median difference −1065, Hodges–Lehmann difference −992.8) and EPI pigs receiving amylase (actual median difference −713.0, Hodges–Lehmann difference

https://doi.org/10.3390/biom16010172

−787.5,) (Figure 3B), possibly increasing absorption and minimising glucose expenditure compared to that observed in the control and microbial amylase groups.

There was a trend ($p < 0.1$) for reduced AUC in EPI pigs during the MMTT with the HFD + SUB compared to that observed during the MMTT with only the HFD (Figure 3D).

A mixed effects analysis was conducted to examine the effects of MMTT meal composition and enzymatic treatment on net postprandial glycemic response. Results showed a significant interaction between MMTT composition and enzymatic treatment ($F(10,38) = 2.1$, $p = 0.0449$). Comparisons of simple effects on net postprandial glycemic response for specific MMTTs in healthy and EPI pigs are presented in Figure 4. The postprandial glycemic response of both healthy control pigs and healthy animals supplemented with amylase was significantly ($p = 0.027$ and 0.007, respectively) influenced by MMTT composition (Figure 4). At the same time, EPI control and EPI amylase animals demonstrated no MMTT composition-dependent changes in net postprandial glycemic response, while EPI Creon animals' postprandial glucose curves were obviously different in all three types of MMTTs ($p = 0.0344$ and 0.0217, for HFD MTT vs. HFD + SUBS MMTT and HFD + SUBS MMTT vs. SUBS MMTT, respectively). EPI pigs receiving amylase, however, demonstrated significantly lower glucose response when compared to the Creon-supplemented group, in HFD and SUBS MMTT (Figure 4). It should be noted that due to the significant interactions between MMTT compositions and enzymatic treatments, no significant main effects were reported. No significant influence of random effect (animal) or its interactions with main effects (MMTTs composition and enzymatic treatment) was observed either.

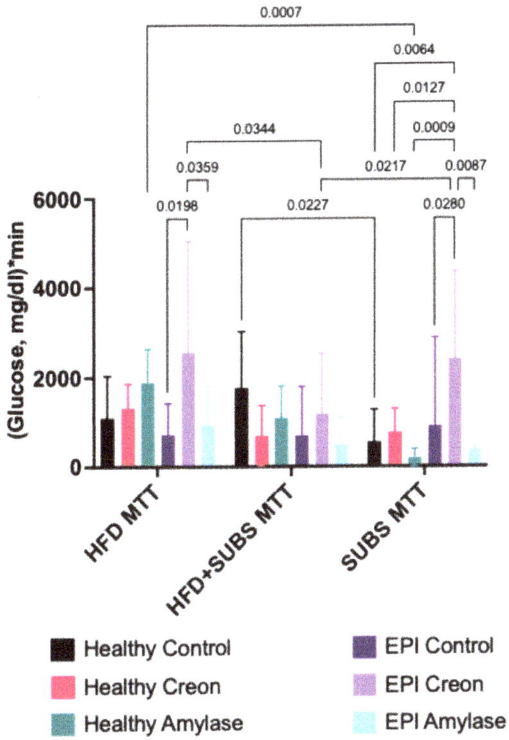

Figure 4. Net postprandial glycemic response AUCs describing the effects of MMTTs composition. Data are presented as mean ± standard deviation (SD) and are baseline adjusted. Differences between the results were considered significant at $p < 0.05$; $p < 0.1$ was considered a trend; p-values are given with the result bars.

https://doi.org/10.3390/biom16010172

Amylase or Creon supplementation in healthy pigs did not affect insulin release (Figure 5A,B). Insulin release in EPI pigs during the HFD + SUB MMTT was lower (Figure 5A,B) than that observed in healthy pigs. Insulin secretion after Creon supplementation to EPI pigs was significantly higher compared to that observed in animals receiving amylase-supplemented feed or in control EPI pigs. The lowest insulin secretion in EPI pigs treated with microbial amylase corresponded to the lowest net postprandial glycemic response during the HFD + SUB MMTT.

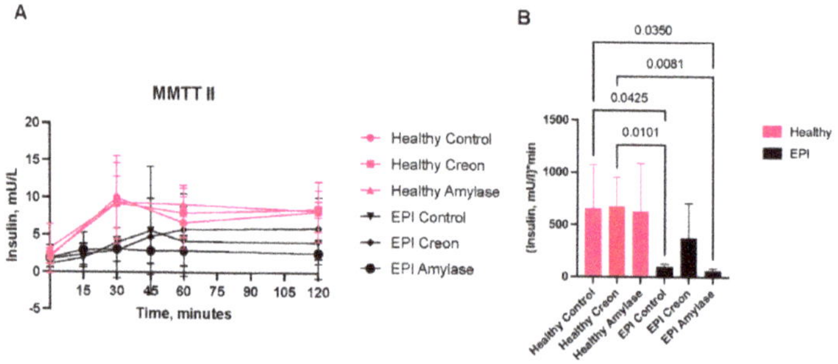

Figure 5. (**A**) Plasma insulin concentrations during mixed meal tolerance test (MMTT) with HFD + SUB; (**B**) comparison of area under the curve (AUC) for insulin release during mixed meal tolerance test (MMTT) with HFD + SUB. Data are expressed as mean ± standard deviation (SD). Area under the curve (AUC) data is baseline adjusted. Differences between the results were considered significant at $p < 0.05$; $p < 0.1$ was considered a trend; p-values are given with the result bars.

4. Discussion

The global prevalence of diabetes mellitus and other metabolic diseases is rapidly growing. According to WHO recommendations, the oral glucose tolerance test (OGTT) is still considered to be a gold standard in the diagnosis of diabetes mellitus (https://www.who.int/news-room/fact-sheets/detail/diabetes (accessed on 9 January 2026)) [26]. However, it is well-known that the OGTT is not a perfect diagnostic tool, as it does not reflect the interactions between nutritional components and their metabolic path during digestion and absorption [27]. The OGTT results in the rapid absorption of glucose, followed by a fast insulin peak, and can cause reactive hypoglycaemia and gastrointestinal symptoms, such as bloating, nausea, and even anxiety in some patients [28,29]. Thus, many patients do not complete the initial OGTT or any follow-up investigations. The need for other measurements, allowing for the successful assessment of postprandial glucose response, is emerging.

To propose an alternative to the OGTT, attempts to introduce continuous glucose monitoring (CGM) [30–32] or even communication-inspired eye diagrams (sub-THz Gluco-Eye) [33] are ongoing. Efforts have also been made to standardize the MMTT and introduce it for use in clinical practice [27,30]. However, the test meal composition, nutritional value, adaptation to its components, and interactions between this meal and the health status of patients have not been analyzed to date.

In the present study, we tried to analyze the influence of MMTT meal composition and health status (enzymatic treatment) in the well-established porcine EPI model. We have demonstrated that postprandial glucose levels after the MMTT in healthy pigs were higher compared to those observed in EPI pigs and thus, must be decreased to physiological values of between 90 and 110 mg/dL for the maintenance of homeostasis (Figure 2). The net postprandial glycemic response curve in healthy pigs was not affected by exogenous enzyme supplementation (Figure 3A); however, it was significantly affected by meal composition and/or adaptation to the feed macronutrient content (Figure 3C).

https://doi.org/10.3390/biom16010172

Following the development of EPI, enzyme supplementation had different modulatory effects on the glucose homeostasis of the pigs. Creon, which is a mixture of many components derived from porcine pancreatic glands (containing enzymes and other biologically active components), significantly increased net postprandial glycemic response (possibly limiting host metabolism and lowering glucose tissue utilization) compared to microbial amylase (Figure 3B). The strongest homeostatic effect of microbial amylase on net postprandial glycemic response or the intensification of glucose expenditure was observed during the SUB MMTT (with unfamiliar macronutrients) in EPI pigs, when compared to the HFD MMTT or HFD + SUB MMTT, containing macronutrients to which the pigs' gastrointestinal tract was already adapted for a long period of time.

In the last century, Mohan et al. [34] showed that after 6 months of Creon treatment, patients achieved better control of diabetes, improvement in abdominal symptoms, and overall sense of well-being. This included a significant reduction in post-prandial plasma glucose and glycosylated hemoglobin at six months vs. baseline [34]. In our study, Creon was found to increase net postprandial glycemic response during the HFD MMTT in EPI pigs compared to pigs treated with microbial amylase. The most controlled glucose AUCs were obtained in EPI pigs treated with microbial amylase, with the highest AUCs obtained in EPI pigs treated with Creon (Figure 4). This confirms that both exposure to MMTT test meal components and microbial amylase treatment interact to modulate postprandial blood glucose homeostasis. As previously mentioned, the separate effects of enzymatic treatment and MMTTs composition were found to be non-significant, but this could probably be explained by the significant interactions between these two factors, which, in turn, is an extremely important point for consideration in the possible clinical use of MMTTs. In the present study, Creon was shown to increase postprandial blood glucose levels and stimulate insulin release, which paradoxically suggests an enhanced insulin resistance. Amylase supplementation in EPI animals led to a decreased net postprandial glucose response, when compared to Creon, thus maintaining insulin sensitivity and preventing pancreatic β-cells from exhaustion. Creon contains active lipase, protease, and amylase, and the difference in effects observed for Creon and one of its main components may be explained by the presence of protease, which was shown to decrease insulin sensitivity [25]. Amylase, in turn, was shown to improve insulin sensitivity and glucose homeostasis [35–37].

Wortham & Sander [38] stated that "Insulin secretion must be tightly coupled to nutritional state to maintain blood glucose homeostasis. To this end, pancreatic β-cells sense and respond to changes in metabolic conditions, thereby anticipating insulin demands for a given physiological context." Interestingly, in our study during the MMTT with the HFD + SUB, the lowest insulin levels corresponded to the lowest glucose AUCs in EPI pigs, during both the control experiment and the experiment with microbial amylase, whereas during the Creon experiment, the secretion of insulin tended to increase following the MMTT (Figure 5B). In healthy pigs, the long-term supplementation of neither Creon nor microbial amylase had any effect on insulin release and the corresponding net postprandial glycemic response. When the effects of the different substrates were compared in relation to microbial amylase vs. Creon supplementation, the results obtained confirmed the above statements.

MMTT meals, composed of new (different from those used for adaptation) types of macronutrients (SUB), definitively altered net postprandial glycemic response and the metabolic expenditure of glucose, downregulating glucose homeostasis. Thus, standardized MMTTs, with test meals composed of macronutrient components unfamiliar to subjects (patients), should be considered as highly relevant tests for obtaining comparable results concerning the precise estimation of glucose homeostasis regulation and insulin release, e.g., for pancreatic enzyme replacement therapy (PERT), metabolic syndrome DT2, etc.

The main limitation of the present study is the lack of opportunity to obtain a complete evaluation of the acute effects (non-adapted) of pancreatic enzyme secretion on glucose homeostasis. The lack of focus on the role of the other enzymes contained within Creon (e.g., lipase and protease) on insulin, incretins, and other gut hormones is another limitation that will be considered in our upcoming experiments. However, a good piece of work related to the above-mentioned limitations has been presented in the article by Pierzynowski et al., 2016 [39].

5. Conclusions

In summary, the pancreatic enzymes, with their possible regulatory features, are less well recognized as regulators of glucose homeostasis, which function in parallel to insulin and incretins, and their role is often neglected. The homeostatic regulation of glucose levels in EPI pigs maintained the net postprandial glycemic response at a low level (probably due to impaired digestive and/or absorption pathways), while glucose expenditure may even be increased compared to healthy animals. Creon supplementation in EPI pigs caused the attenuation of glucose homeostasis regulatory mechanisms, increasing the net postprandial glycemic response and probably leading to enhanced insulin resistance, which in turn, limits the metabolic expenditure of glucose. In contrast to Creon, microbial amylase-dependent regulatory mechanisms of glucose homeostasis maintenance decreased the net postprandial glycemic response and eventually increased the metabolic expenditure of glucose. Thus, microbial amylase exhibits both insulin-like and anti-incretin features.

Author Contributions: Conceptualization, P.W., S.G.P. and K.P.; methodology K.Z. and K.P.; validation, K.P., R.G. and P.W.; formal analysis, K.Z. and D.S.; investigation, J.W., T.J., D.S. and K.Z.; resources, P.W., S.G.P., K.P. and T.J.; data curation, K.P. and K.Z.; writing—original draft preparation, P.W., S.G.P. and K.P.; writing—review and editing, J.D.; visualization, K.P. and T.J.; supervision, S.G.P.; project administration, K.P.; funding acquisition, P.W. and S.G.P. All authors have read and agreed to the published version of the manuscript.

Funding: This research was funded by Anara AB (Sweden), MED Sp z o.o. (Poland), National Science Centre (Poland) (Preludium, 2021/41/N/NZ9/01722) and Specialized Private Implantology Clinic Wychowanski Stomatologia.

Institutional Review Board Statement: The animal study protocol was approved by the Second Local Ethics Committee for Animal Experimentation in Warsaw, Poland (approval no. WAW2/025/2022, approval date 16 February 2022).

Informed Consent Statement: Not applicable.

Data Availability Statement: The original contributions presented in this study are included in the article. Further inquiries can be directed to the corresponding authors.

Conflicts of Interest: The authors of this manuscript have the following competing interests: Stefan Pierzynowski and Kateryna Pierzynowska are the owners of Anara AB, Sweden. Janine Donaldson is employed by Anara AB, Sweden. Piotr Wychowanski is the owner of the Specialized Private Implantology Clinic Wychowanski Stomatologia and MED Sp z o.o. (Poland), Robert Gallotto is Founder and CEO of Anagram Therapeutics, Inc. (USA). All other authors declare no competing interests.

Abbreviations

The following abbreviations are used in this manuscript:

AUC	Area under the curve
CF	Cystic fibrosis
DT2	Diabetes type 2
EPI	Exocrine pancreatic insufficiency

https://doi.org/10.3390/biom16010172

HFD	High-fat diet
MMTT	Mixed meal tolerance test
OGTT	Oral glucose tolerance test
PDL	Pancreatic duct ligation
PERT	Pancreatic enzyme replacement therapy
SUB	Substrate

References

1. Jin, S.; Sha, L.; Dong, J.; Yi, J.; Liu, Y.; Guo, Z.; Hu, B. Effects of Nutritional Strategies on Glucose Homeostasis in Gestational Diabetes Mellitus: A Systematic Review and Network Meta-Analysis. *J. Diabetes Res.* **2020**, 6062478. [CrossRef]
2. Mitchell, C.M.; Stinson, E.J.; Chang, D.C.; Krakoff, J. A Mixed Meal Tolerance Test Predicts Onset of Type 2 Diabetes in Southwestern Indigenous Adults. *Nutr. Diabetes* **2024**, *14*, 50. [CrossRef]
3. Gastaldelli, A. Measuring and Estimating Insulin Resistance in Clinical and Research Settings. *Obesity* **2022**, *30*, 1549–1563. [CrossRef]
4. Röder, P.V.; Wu, B.; Liu, Y.; Han, W. Pancreatic Regulation of Glucose Homeostasis. *Exp. Mol. Med.* **2016**, *48*, e219. [CrossRef]
5. Tirone, T.A.; Brunicardi, F.C. Overview of Glucose Regulation. *World J. Surg.* **2001**, *25*, 461–467. [CrossRef] [PubMed]
6. van Baak, M.A.; Roumans, N.J.T.; Mariman, E.C.M. Diet Composition, Glucose Homeostasis, and Weight Regain in the YoYo Study. *Nutrients* **2021**, *13*, 2257. [CrossRef]
7. Leppälä, J.; Lagström, H.; Kaljonen, A.; Laitinen, K. Construction and Evaluation of a Self-Contained Index for Assessment of Diet Quality. *Scand. J. Public Health* **2010**, *38*, 794–802. [CrossRef] [PubMed]
8. Guasch-Ferré, M.; Willett, W.C. The Mediterranean Diet and Health: A Comprehensive Overview. *J. Intern. Med.* **2021**, *290*, 549–566. [CrossRef]
9. Castro-Barquero, S.; Ruiz-León, A.M.; Sierra-Pérez, M.; Estruch, R.; Casas, R. Dietary Strategies for Metabolic Syndrome: A Comprehensive Review. *Nutrients* **2020**, *12*, 2983. [CrossRef]
10. Jackson, A.D.; McLaughlin, J. Digestion and Absorption. *Surgery* **2009**, *27*, 231–236. [CrossRef]
11. Normatov, I.; Sentongo, T. Pancreatic Malnutrition in Children. *Pediatr. Ann.* **2019**, *48*, e441–e447. [CrossRef]
12. Layer, P.; Keller, J. Pancreatic Enzymes: Secretion and Luminal Nutrient Digestion in Health and Disease. *J. Clin. Gastroenterol.* **1999**, *28*, 3–10. [CrossRef]
13. Hou, N.; Du, X.; Wu, S. Advances in Pig Models of Human Diseases. *Anim. Models Exp. Med.* **2022**, *5*, 141–152. [CrossRef]
14. Lunney, J.K.; Van Goor, A.; Walker, K.E.; Hailstock, T.; Franklin, J.; Dai, C. Importance of the Pig as a Human Biomedical Model. *Sci. Transl. Med.* **2021**, *13*, eabd5758. [CrossRef] [PubMed]
15. Roura, E.; Koopmans, S.-J.; Lallès, J.-P.; Le Huerou-Luron, I.; De Jager, N.; Schuurman, T.; Val-Laillet, D. Critical Review Evaluating the Pig as a Model for Human Nutritional Physiology. *Nutr. Res. Rev.* **2016**, *29*, 60–90. [CrossRef]
16. Pierzynowska, K.; Wychowański, P.; Zaworski, K.; Woliński, J.; Donaldson, J.; Pierzynowski, S. Anti-Incretin Gut Features Induced by Feed Supplementation with Alpha-Amylase: Studies on EPI Pigs. *Int. J. Mol. Sci.* **2023**, *24*, 16177. [CrossRef] [PubMed]
17. Pierzynowski, S.G.; Stier, C.; Pierzynowska, K. Hypothesis That Alpha-Amylase Evokes Regulatory Mechanisms Originating in the Pancreas, Gut and Circulation, Which Govern Glucose/Insulin Homeostasis. *World J. Diabetes* **2023**, *14*, 1341–1348. [CrossRef]
18. Pierzynowska, K.G.; Lozinska, L.; Woliński, J.; Pierzynowski, S. The Inverse Relationship between Blood Amylase and Insulin Levels in Pigs during Development, Bariatric Surgery, and Intravenous Infusion of Amylase. *PLoS ONE* **2018**, *13*, e0198672. [CrossRef] [PubMed]
19. Faul, F.; Erdfelder, E.; Lang, A.-G.; Buchner, A. G*Power 3: A Flexible Statistical Power Analysis Program for the Social, Behavioral, and Biomedical Sciences. *Behav. Res. Methods* **2007**, *39*, 175–191. [CrossRef]
20. Donaldson, J.; Fed'kiv, O.; Pawłowska, M.; Kowalik, S.; Erlwanger, K.H.; Weström, B.; Kruszewska, D.; Pierzynowski, S.G. The Effectiveness of Enzymatic Replacement Therapy Measured by Turbidimetry and the Lipaemic Index in Exocrine Pancreatic Insufficient Young, Growing Pigs, Fed a High-Fat Diet. *Adv. Med. Sci.* **2009**, *54*, 7–13. [CrossRef] [PubMed]
21. Freedman, S.D.; Zaworski, K.; Pierzynowska, K.; Pierzynowski, S.; Gallotto, R.; Sathe, M.; Borowitz, D.S. Validation of an Omega-3 Substrate Challenge Absorption Test as an Indicator of Global Fat Lipolysis. *PLoS ONE* **2023**, *18*, e0284651. [CrossRef]
22. *Nutrient Requirements of Swine/Committee on Nutrient Requirements of Swine*, 11th rev. ed.; Board on Agriculture and Natural Resources, Division on Earth and Life Studies; The National Academies Press: Washington, DC, USA, 2012.
23. Szczurek, P.; Mosiichuk, N.; Woliński, J.; Yatsenko, T.; Grujic, D.; Lozinska, L.; Pieszka, M.; Święch, E.; Pierzynowski, S.G.; Goncharova, K. Oral Uricase Eliminates Blood Uric Acid in the Hyperuricemic Pig Model. *PLoS ONE* **2017**, *12*, e0179195. [CrossRef]

https://doi.org/10.3390/biom16010172

24. Pierzynowska, K.; Pierzynowski, S.; Zaworski, K.; Gallotto, R.; Sathe, M.; Freedman, S.D.; Borowitz, D. Validation of Dietary Whey Protein Substrate Challenge Absorption Test as an Indicator of Proteolysis. *PLoS ONE* **2025**, *20*, e0323730. [CrossRef]
25. Althouse, A.D. Adjust for Multiple Comparisons? It's Not That Simple. *Ann. Thorac. Surg.* **2016**, *101*, 1644–1645. [CrossRef]
26. Phillips, P.J. Oral Glucose Tolerance Testing. *Aust. Fam. Physician* **2012**, *41*, 391–393.
27. Lages, M.; Barros, R.; Moreira, P.; Guarino, M.P. Metabolic Effects of an Oral Glucose Tolerance Test Compared to the Mixed Meal Tolerance Tests: A Narrative Review. *Nutrients* **2022**, *14*, 2032. [CrossRef] [PubMed]
28. Brodovicz, K.G.; Girman, C.J.; Simonis-Bik, A.M.C.; Rijkelijkhuizen, J.M.; Zelis, M.; Bunck, M.C.; Mari, A.; Nijpels, G.; Eekhoff, E.M.W.; Dekker, J.M. Postprandial Metabolic Responses to Mixed versus Liquid Meal Tests in Healthy Men and Men with Type 2 Diabetes. *Diabetes Res. Clin. Pract.* **2011**, *94*, 449–455. [CrossRef]
29. Hillier, T.A.; Vesco, K.K.; Pedula, K.L.; Beil, T.L.; Whitlock, E.P.; Pettitt, D.J. Screening for Gestational Diabetes Mellitus: A Systematic Review for the U.S. Preventive Services Task Force. *Ann. Intern. Med.* **2008**, *148*, 766–775. [CrossRef] [PubMed]
30. Ramos-Levi, A.M.; Rubio-Herrera, M.A.; Matía-Martín, P.; Pérez-Ferre, N.; Marcuello, C.; Sánchez-Pernaute, A.; Torres-García, A.J.; Calle-Pascual, A.L. Mixed Meal Tolerance Test Versus Continuous Glucose Monitoring for an Effective Diagnosis of Persistent Post-Bariatric Hypoglycemia. *JCM* **2023**, *12*, 4295. [CrossRef]
31. Ferreira, R.O.M.; Trevisan, T.; Pasqualotto, E.; Chavez, M.P.; Marques, B.F.; Lamounier, R.N.; Van De Sande-Lee, S. Continuous Glucose Monitoring Systems in Noninsulin-Treated People with Type 2 Diabetes: A Systematic Review and Meta-Analysis of Randomized Controlled Trials. *Diabetes Technol. Ther.* **2024**, *26*, 252–262. [CrossRef]
32. Ajjan, R.A.; Battelino, T.; Cos, X.; Del Prato, S.; Philips, J.-C.; Meyer, L.; Seufert, J.; Seidu, S. Continuous Glucose Monitoring for the Routine Care of Type 2 Diabetes Mellitus. *Nat. Rev. Endocrinol.* **2024**, *20*, 426–440. [CrossRef] [PubMed]
33. Lee, W.; Kim, E.S.; Kim, S.; Park, H.; Lee, J.K.; Baek, E.; Jung, C.H.; Cho, S.-H. Noninvasive Subterahertz Glucose Monitoring Using a Communication Inspired Eye Diagram. *Sci. Rep.* **2025**, *16*, 1079. [CrossRef]
34. Mohan, V.; Poongothai, S.; Pitchumoni, C.S. Oral Pancreatic Enzyme Therapy in the Control of Diabetes Mellitus in Tropical Calculous Pancreatitis. *Int. J. Pancreatol.* **1998**, *24*, 19–22. [CrossRef] [PubMed]
35. Date, K.; Satoh, A.; Iida, K.; Ogawa, H. Pancreatic α-Amylase Controls Glucose Assimilation by Duodenal Retrieval through N-Glycan-Specific Binding, Endocytosis, and Degradation. *J. Biol. Chem.* **2015**, *290*, 17439–17450. [CrossRef]
36. Asanuma-Date, K.; Hirano, Y.; Le, N.; Sano, K.; Kawasaki, N.; Hashii, N.; Hiruta, Y.; Nakayama, K.; Umemura, M.; Ishikawa, K.; et al. Functional Regulation of Sugar Assimilation by N-Glycan-Specific Interaction of Pancreatic α-Amylase with Glycoproteins of Duodenal Brush Border Membrane. *J. Biol. Chem.* **2012**, *287*, 23104–23118. [CrossRef]
37. Bonnefond, A.; Yengo, L.; Dechaume, A.; Canouil, M.; Castelain, M.; Roger, E.; Allegaert, F.; Caiazzo, R.; Raverdy, V.; Pigeyre, M.; et al. Relationship between Salivary/Pancreatic Amylase and Body Mass Index: A Systems Biology Approach. *BMC Med.* **2017**, *15*, 37. [CrossRef] [PubMed]
38. Wortham, M.; Sander, M. Mechanisms of B-cell Functional Adaptation to Changes in Workload. *Diabetes Obes. Metab.* **2016**, *18*, 78–86. [CrossRef]
39. Pierzynowski, S.; Goncharova, K.; Woliński, J.; Prykhodko, O.; Weström, B.; Lozinska, L. Enteral Pancreatic-like Enzymes of Microbial Origin Affect Insulin Release during an Intravenous Glucose Tolerance Test. *J. Diabetes Metab.* **2016**, *7*, 681. [CrossRef]

https://doi.org/10.3390/biom16010172

biomolecules

MDPI

Article

Feedback Regulation of Pancreatic Juice Secretion in Pigs

Jose Luis Valverde Piedra [1,*] and Sylwia Edyta Szymanczyk [2]

1 Department of Pharmacology, Toxicology and Environmental Protection, Faculty of Veterinary Medicine, University of Life Sciences in Lublin, Akademicka 12, 20-033 Lublin, Poland
2 Department of Animal Physiology, Faculty of Veterinary Medicine, University of Life Sciences in Lublin, Akademicka 12, 20-950 Lublin, Poland; sylwia.szymanczyk@up.edu.pl
* Correspondence: jose.valverde@up.edu.pl

Abstract

Pancreatic exocrine secretion is regulated by the physicochemical properties and nutrient composition of gastric and intestinal chyme. The present study examined integrative feedback mechanisms involved in the physiological control of pancreatic secretion, with particular emphasis on interactions between pancreatic juice, bile, and gut-derived regulatory and metabolic signals. A chronic porcine model enabling selective withdrawal and controlled reintroduction of pancreatic juice and bile into defined intestinal segments was employed. Duodenal and ileal exposure to pancreatic juice suppressed pancreatic enzyme secretion, while intraduodenal administration of pancreatin elicited a biphasic inhibitory response. Interruption of bile flow to the duodenum resulted in increased pancreatic protein output and was associated with reduced circulating cholecystokinin concentrations. In contrast, intraduodenal infusion of bile acids attenuated pancreatic exocrine secretion. Prolonged bile deprivation led to sustained pancreatic hypersecretion accompanied by a marked reduction in biliary leptin output. Collectively, these findings indicate that pancreatic exocrine secretion in pigs is regulated by multiple interacting feedback pathways operating along the gastrointestinal tract. The observed responses support functional contributions of protease-dependent luminal feedback, distal intestinal sensing, hormone-dependent regulation, and bile-associated metabolic modulation.

Keywords: pancreatic secretion; negative feedback; bile acids; CCK; leptin; ileal brake; FXR/TGR5

1. Introduction

Exocrine pancreatic secretion is regulated by an integrated network of neural and hormonal mechanisms that adjust the delivery of digestive enzymes and bicarbonate to the composition and timing of gastric and intestinal chyme [1]. Although these regulatory pathways have been extensively studied, feedback mechanisms remain incompletely translated across species. This variability reflects differences in the relative contribution of CCK-dependent acinar stimulation, vagal reflex pathways, and ductal secretin signalling between rodents, large mammals, and humans [2].

The pig represents an established translational model for gastrointestinal physiology and nutritional research [3]. However, in this species, several aspects of pancreatic feedback regulation remain insufficiently resolved, particularly those related to bile components and distal intestinal sensing. One of the central elements of feedback control is protease-dependent modulation of CCK release. In humans, intraduodenal trypsin reduces circulating CCK concentrations, demonstrating that luminal proteolytic activity can down-regulate pancreatic enzyme secretion through hormonal pathways [4]. Rodent studies further identified trypsin-sensitive CCK-releasing peptides, including luminal CCK-releasing factor and monitor peptide, which accumulate when protease activity is reduced or functionally consumed by dietary substrates, thereby enhancing I-cell stimulation and pancreatic enzyme output [5–7]. Despite these advances, the range of luminal protease activity over which this feedback operates, as well as its saturation characteristics, are not well defined in pigs under chronic feeding conditions. It also remains unclear whether this regulation is restricted to the duodenum or extends to more distal intestinal segments. In large mammals, CCK-induced pancreatic secretion is often mediated predominantly through neural amplification, whereby CCK1 receptors on vagal afferents initiate vago-vagal reflexes that enhance cholinergic stimulation of acinar cells [2]. At the cellular level, CCK activates phospholipase C–dependent Ca^{2+} signalling in acinar tissue [8]. Importantly, porcine pancreatic tissue displays differences in CCK receptor expression and functional responsiveness compared with rodents, which complicates direct extrapolation of rodent-derived models [9]. In parallel, secretin remains the principal regulator of bicarbonate-rich ductal secretion via cAMP-dependent mechanisms [10]. Physiological pancreatic output therefore emerges from coordinated interactions between multiple regulatory pathways rather than from a single dominant secretagogue.

Beyond the classical protease–CCK axis, accumulating evidence indicates that bile and bile acids influence pancreatic exocrine secretion. However, the direction and magnitude of these effects vary considerably between experimental models. One contributing factor may be the highly dynamic nature of intestinal bile acid concentrations, which fluctuate markedly following gallbladder emptying and reach millimolar levels in the proximal intestine after feeding in humans, while systemic concentrations remain substantially lower [11,12]. These gradients enable bile acids to function as local intestinal signals capable of modulating enteroendocrine activity and neural reflexes rather than acting solely as detergents for lipid digestion. Two bile acid-responsive pathways are of particular interest in this context. Farnesoid X receptor (FXR) links intestinal bile acid exposure to endocrine feedback through FGF15/19 signalling [13], whereas the G protein–coupled receptor TGR5 (GPBAR1) increases intracellular cAMP and promotes incretin release, including GLP-1, thereby influencing gastrointestinal secretory regulation [14]. Although these pathways are well characterised in metabolic physiology, their integration with in vivo pancreatic enzyme feedback remains insufficiently examined, especially in large-animal models allowing independent manipulation of bile and pancreatic juice.

Importantly, intestinal feedback regulation may not be confined to the proximal small intestine. In pigs, luminal delivery of pancreatic juice or bile to different intestinal segments has been reported to modify pancreatic secretory output, suggesting that distal intestinal sensing mechanisms resembling an "ileal brake" may contribute to overall regulation [15,16].

This concept is physiologically plausible, as nutrient and bile acid detection in the ileum could modulate pancreatic secretion once digestion progresses distally, thereby limiting energetically inefficient enzyme oversupply [2]. Nevertheless, the relative contribution of bile acids and pancreatic enzymes to such distal feedback, as well as the hormonal mediators involved, remains unclear in weaned pigs studied under chronic cannulation conditions [17].

Finally, pancreatic exocrine secretion may also be influenced by metabolic hormones traditionally associated with appetite and energy homeostasis. Leptin has been reported to modulate pancreatic secretion and to interact with CCK-dependent pathways in a context- and dose-dependent manner in experimental models [18,19]. However, the potential role of biliary or intestinal leptin in phase-dependent fine-tuning of pancreatic exocrine output

https://doi.org/10.3390/biom16020322

remains poorly characterised in large mammals. This limitation largely reflects the scarcity of studies combining direct measurements of bile composition with pancreatic secretory responses in conscious, chronically instrumented animals [17,20].

Aim of the Study

The aim of this study was to investigate the relationships between four conceptual regulatory levels involved in pancreatic exocrine control (Figure 1). These levels comprised (i) luminal pancreatic juice and enzyme delivery as a primary protease-dependent feedback signal, (ii) hormone-dependent pathways with particular emphasis on cholecystokinin (CCK) and acid/secretin-dependent mechanisms, (iii) bile delivery and bile acid exposure as modulatory luminal and enteroendocrine signals, and (iv) bile-associated metabolic signalling, with specific focus on leptin as a potential phase-dependent modulatory factor. The study was conducted in weaned piglets equipped with chronic pancreatic and bile duct catheters and duodenal and ileal access.

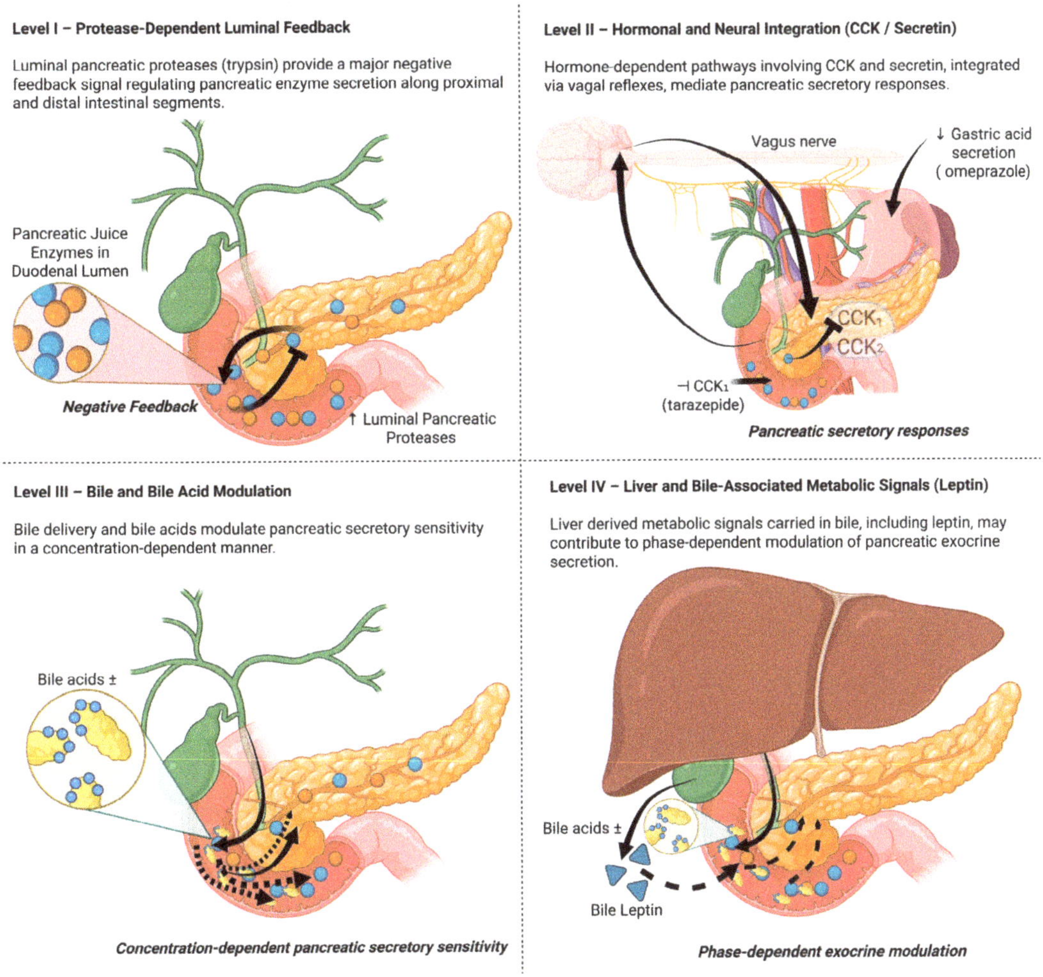

Figure 1. Conceptual framework of multilevel feedback regulation of pancreatic exocrine secretion in pigs. The scheme illustrates regulatory pathways investigated in the present study and does not imply exclusive or strictly sequential control. Created using BioRender (https://www.biorender.com/) [21].

https://doi.org/10.3390/biom16020322

In the present work, pancreatic exocrine regulation is conceptualised as a system of hierarchically organised feedback loops, defined as a multilevel control network in which pancreatic secretion is regulated sequentially and in parallel by direct luminal signals, hormone-dependent pathways, and bile- and metabolism-related modulatory mechanisms.

To investigate these relationships, two complementary chronic porcine experimental models were employed.

Using Experimental Model I, we tested the hypothesis that luminal delivery of pancreatic enzymes constitutes the primary inhibitory feedback mechanism controlling pancreatic exocrine secretion. Specifically, we hypothesised that graded intraluminal administration of pancreatic enzymes suppresses pancreatic secretion in a dose- and site-dependent manner along both proximal (duodenal) and distal (ileal) segments of the small intestine, including conditions of interrupted endogenous pancreatic juice flow.

Using Experimental Model II, we tested the hypothesis that bile delivery and bile acid exposure represent independent modulatory signals in the regulation of pancreatic exocrine secretion and that interruption of bile flow results in sustained stimulation of pancreatic enzyme output. This model further enabled evaluation of the contribution of CCK- and acid/secretin-dependent pathways through pharmacological blockade of CCK_1 receptors and inhibition of gastric acid-stimulated secretin release, as well as assessment of bile-borne leptin as a potential metabolic signal associated with pancreatic secretory regulation.

By integrating selective withdrawal and controlled reintroduction of pancreatic juice and bile with targeted pharmacological interventions, this study aimed to delineate the relative contribution and functional hierarchy of luminal enzyme feedback, bile acid signalling, hormonal pathways, and bile-associated metabolic cues in the coordinated regulation of pancreatic exocrine function.

2. Materials and Methods

The study was designed to evaluate the regulatory mechanisms involved in feedback control of pancreatic exocrine secretion under controlled experimental conditions.

2.1. Animals

The experiments were performed on weaned piglets of a commercial crossbreed (Large White Swedish × Hampshire × Yorkshire) with an initial body mass of 15 ± 5 kg. Animals were obtained from the experimental farm in Odarslöv belonging to the Swedish University of Agricultural Sciences (Alnarp, Sweden).

All animal procedures were approved by the Local Ethics Committee of Lund University, Sweden (permission no. M238-99).

2.2. Preparation of Experimental Model I

2.2.1. Surgical Procedures

The relationship between the activity of pancreatic proteolytic enzymes in the small intestine and pancreatic juice secretion was achieved using the model described by Pierzynowski et al., Thaela et al., and Rantzer et al. [15,22–24].

Piglets were premedicated with azaperone (2 mg/kg body weight; Stresnil, Janssen Pharmaceutica, Beerse, Belgium), and general anaesthesia was induced with halothane (3% in air; ISC Chemicals Ltd., Avonmouth, Bristol, UK). All surgical procedures were performed under aseptic conditions.

Animals were surgically fitted with a chronic pancreatic duct catheter and a T-shaped duodenal cannula made of medical-grade silicone, according to the original technique described by Pierzynowski et al. and subsequently modified by Thaela et al. [22,23]. A

https://doi.org/10.3390/biom16020322

catheter was additionally placed in the external jugular vein to allow stress-free blood sampling during experimental procedures.

The pancreatic duct catheter enabled continuous collection of pancreatic juice, whereas the duodenal cannula allowed free return of pancreatic secretion to the intestinal lumen under control conditions. During experimental interventions, this cannula permitted controlled intraduodenal administration of pancreatic juice or test solutions. An additional catheter inserted into the ileum provided direct access to the distal small intestine for site-specific delivery of pancreatic enzymes.

2.2.2. Postoperative Care and Animal Maintenance

After surgery, piglets were housed individually in pens with free access to drinking water. Visual contact between animals was maintained to minimise social isolation-related stress.

From the day of surgery until the end of the experimental period, animals were monitored daily. Clinical assessment included measurement of rectal temperature, heart rate, respiratory rate, general health status, and regular recording of body weight. Experimental procedures were initiated after a minimum recovery period of seven days, when piglets were clinically stable and exhibited normal feeding behaviour.

For five days following the surgical intervention, piglets received intramuscular antibiotic therapy. Postoperative analgesia was provided according to institutional veterinary recommendations, and animals were continuously monitored for signs of pain, infection, catheter-related complications, or reduced feed intake. Predefined humane endpoints included sustained weight loss, signs of systemic infection, persistent catheter occlusion, or behavioural indicators of distress.

To compensate for water and electrolyte losses resulting from temporary diversion of pancreatic juice from the duodenum, animals were provided with an oral mineral supplement (Salt Balance, Lantmännen, Stockholm, Sweden) throughout the recovery period and after each experimental procedure, in accordance with the manufacturer's guidelines.

The established experimental models required continuous 24 h monitoring. Patency of catheters and duodenal cannulas was evaluated several times daily. These inspections were essential to prevent obstruction of pancreatic and biliary flow into the duodenum and to mitigate the risk of pancreatitis and cholecystitis. Particular attention was paid to the care of the skin surrounding the cervical catheter site, the abdominal integument, and the limbs and hooves, with the aim of reducing pruritus around healing wounds and minimising irritation caused by pancreatic enzymes. For this purpose, a topical emollient (Kylbalsam, ACO Hud AB, Stockholm, Sweden) and a protective cream against enzymatic irritation (Helosan, Pharmacia Animal Health AB, Stockholm, Sweden) were applied at least twice daily.

Animals were fed twice daily with a standard commercial pig feed mixture (Växfor, Lantmännen, Sweden; Table 1), administered in the morning (10:00–11:00) and afternoon (15:30–16:30). Feed portions (180–350 g), corresponding to approximately 1% of body mass, were consumed within one hour under both control and experimental conditions. Throughout the study, piglets had continuous access to drinking water.

https://doi.org/10.3390/biom16020322

Table 1. Composition of the commercial feed mixture for piglets used in the experiments.

Composition	Content
Metabolizable Energy (Mj·Kg^{-1})	12.2
Total Protein (%)	15.5
Fat (%)	4.5
Fibre (%)	5.0
Lysine (G·Kg^{-1})	9.9
Methionine (G·Kg^{-1})	3.6
Threonine (G·Kg^{-1})	5.9
Calcium (%)	0.75
Phosphorus (%)	0.55
Vitamin A (Iu)	8000
Vitamin D3 (Iu)	800

2.2.3. Experiments Using Model I

To evaluate the role of pancreatic juice in the small intestine, as well as the influence of the amount and site of pancreatic enzyme delivery on exocrine pancreatic function, a total of fourteen experimental protocols were performed using Model I. An overview of all experimental variants is provided in Appendix A.1 (Table A1), while the experimental design is illustrated in Appendix A.2 (Scheme A1).

In the control experiment (Experiment 1), pancreatic juice secretion was assessed during the perifeeding period under conditions of continuous return of pancreatic juice to the duodenum.

In Experiments 2–4, pancreatin solutions with different enzymatic activities (20%, 100%, and 200% of physiological trypsin activity) were infused intraduodenally while pancreatic juice flow to the duodenum was preserved.

In Experiment 5, pancreatic juice flow to the duodenum was completely interrupted.

In Experiments 6–8, pancreatin solutions with different enzymatic activities (20%, 100%, and 200%) were administered intraduodenally under conditions of eliminated pancreatic juice flow.

In Experiments 9–11, pancreatin solutions were infused into the ileum while pancreatic juice flow to the duodenum was preserved.

In Experiments 12–14, ileal pancreatin infusions were performed under conditions of disrupted pancreatic juice flow to the duodenum.

The complete experimental scheme is summarised in Table A1 (Appendix A.1).

2.2.4. Collection of Pancreatic Juice

Experiments were conducted daily and lasted 3 h, comprising a 1 h pre-feeding period (preprandial phase), a 1 h feeding period (prandial phase), and a 1 h post-feeding period (postprandial phase).

Before the start of each experiment, the pancreatic catheter was disconnected from the duodenal cannula for 30 min. During this period, pancreatic juice was collected and subsequently reintroduced into the duodenum using a syringe in small aliquots during the first 30 min of the experimental session.

Pancreatic juice collection during experiments was performed under conditions allowing complete freedom of movement for the piglets. Each experimental phase was divided into two consecutive 30 min collection intervals. From each interval, a 0.5 mL sample of pancreatic juice was collected for subsequent biochemical analyses. The remaining volume was returned to the duodenum during the following 30 min in multiple small portions (2–6 mL), depending on the amount of secreted pancreatic juice.

https://doi.org/10.3390/biom16020322

2.2.5. Pancreatin Infusions

Pancreatin infusions (Pancreatin, Solvay Pharmaceuticals GmbH, Hannover, Germany) were administered intraduodenally or intraileally starting immediately after the onset of feed intake and continued throughout the feeding (prandial) phase.

The reference trypsin activity defined as 100% corresponded to the mean enzymatic activity of pancreatic juice collected from piglets of the same age during the feeding phase under identical dietary conditions. Based on this reference, the pancreatin solution corresponding to 100% trypsin activity (3500 U trypsin) was prepared by dissolving 1.2 g of pancreatin in 24 mL of cooled physiological saline (0.9% NaCl).

Pancreatin solutions corresponding to 20% and 200% of physiological trypsin activity were prepared by dissolving 0.24 g and 2.4 g of pancreatin, respectively, in 24 mL of cooled physiological saline.

2.3. Preparation of Experimental Model II

To investigate the effect of the presence of bile in the small intestine on pancreatic juice secretion, the research model developed by Valverde Piedra and Pierzynowski, allowing for the collection and reintroduction of pancreatic juice or both pancreatic juice and bile simultaneously in the pig, was used [16].

2.3.1. Experimental Design and Treatment Schedule

The study was conducted on six weaned piglets (n = 6) assigned to all treatments according to a Latin square, repeated-measures design.

Under control conditions (B+, PJ+), both pancreatic juice (PJ) and bile (B) were returned to the duodenum. In the experimental conditions, pancreatic juice was returned while bile was diverted (B−, PJ+), or both bile and pancreatic juice were diverted (B−, PJ−).

The experimental schemes for Experiments 15–21 and 22–26 are presented in Appendix A.1 (Tables A2 and A3), and the corresponding protocol designs are illustrated in Appendix A.2 (Schemes A2 and A3).

Collection of pancreatic juice and bile was performed during the morning feeding session and lasted 5 h, comprising a 1 h preprandial period, a 1 h prandial period, and a 3 h postprandial period. Secretions were collected at 30 min intervals; volumes were recorded, and 1.5 mL aliquots were stored at −20 °C for subsequent analyses. The remaining pancreatic juice and bile were either reintroduced into the duodenum during the subsequent collection interval or chilled and reintroduced after completion of the experiments.

2.3.2. Experiments Using Model II

To investigate the contribution of CCK1 receptor–mediated signalling and acid-dependent secretin release to feedback regulation of pancreatic exocrine secretion and bile output, experiments were performed using tarazepide (Solvay Pharmaceuticals GmbH, Hannover, Germany) omeprazole (Losec, AstraZeneca, Södertälje, Sweden), and bicarbonate for buffering of duodenal contents.

The experiments were conducted on six piglets surgically prepared according to Model II. Animals were additionally equipped with a catheter placed in the external jugular vein to allow intravenous administration of omeprazole. The experimental protocols followed the design outlined in Appendix A.1 (Table A2) and Appendix A.2 (Scheme A2).

Each experimental session lasted 3 h and included a preprandial, prandial, and postprandial phase. Body weight was measured before each experiment to calculate individual doses of omeprazole and tarazepide.

In control experiments (Experiment 15), pancreatic juice and bile secretion were measured under conditions of preserved return of both secretions to the duodenum. The effects

https://doi.org/10.3390/biom16020322

of bile diversion alone (Experiment 16) and combined diversion of bile and pancreatic juice (Experiment 17) on pancreatic and biliary secretion were subsequently examined.

Pharmacological interventions were performed under conditions of preserved bile and pancreatic juice flow to the duodenum and included intravenous omeprazole administration (1 mg/kg body mass; Experiment 18), intraduodenal tarazepide administration (5 mg/kg body mass; Experiment 19), and combined administration of omeprazole and tarazepide at the same doses (Experiment 20).

In addition, pancreatic juice secretion was assessed under conditions of eliminated bile flow combined with intraduodenal bicarbonate infusion, administered in volumes corresponding to the bile output measured in individual collection intervals (Experiment 21). The doses of omeprazole and tarazepide were selected based on previously published studies [25–27].

2.3.3. Collection of Pancreatic Juice and Bile

All experiments followed a 3 h experimental protocol, comprising a 1 h preprandial phase, a 1 h prandial phase, and a 1 h postprandial phase. Pancreatic juice and bile were collected in 30 min intervals. At the end of each collection interval, the volume of each secretion was measured, and a 1 mL aliquot was collected for subsequent biochemical analyses. In experiments involving the return of pancreatic juice and bile to the intestine, the collected secretions were reintroduced into the duodenum in small aliquots during the subsequent 30 min interval to mimic physiological flow patterns.

For pharmacological interventions, tarazepide was prepared *ex tempore* as a suspension in 5 mL of bile previously collected from the same piglet and administered intraduodenally at a dose of 5 mg/kg body mass, 90 min before the onset of feed intake. Omeprazole was administered intravenously at a dose of 1.0 mg/kg body mass, 30 min before feeding. The detailed experimental timeline and intervention sequence are illustrated in Appendix A.2 (Scheme A2).

2.3.4. Experiments with 10 h Elimination of Bile Flow to the Duodenum

To assess the role of bile in the regulation of pancreatic exocrine secretion, experiments were performed on six piglets surgically prepared according to Experimental Model II.

In the control condition (Experiment 22; Appendix A.1, Table A3), both pancreatic juice and bile were returned to the duodenum overnight, and disconnection of the pancreatic and bile catheters occurred only during the experimental observation period.

In subsequent experiments (Experiments 23–26), bile flow to the duodenum was eliminated for 10 h prior to the experiment. In Experiment 23, the effect of prolonged bile diversion (10 h before the experiment and 5 h during the experimental period) on pancreatic juice secretion was evaluated.

In Experiments 24–26, pancreatic juice secretion was assessed under conditions of eliminated bile flow during intraduodenal infusion of exogenous bile acid salt solutions. Three concentrations of bile acid salts were tested: 0.02 M (Experiment 24), 0.04 M (Experiment 25), and 0.08 M (Experiment 26), administered at a constant rate of 7.5 mL/kg body mass/hour (Appendix A.1, Table A3).

2.3.5. Experimental Scheme and Bile Acid Salt Infusions Under Interrupted Bile Flow

The experimental scheme for studies with preserved or interrupted bile flow to the duodenum (Experiments 22–26) is summarised in Appendix A.1 (Table A3), and the detailed protocol design is illustrated in Appendix A.2 (Scheme A3).

All experiments lasted 5 h and comprised a 1 h pre-feeding phase, a 1 h feeding phase, and a 3 h post-feeding observation period, conducted under conditions of either preserved or eliminated bile flow to the duodenum.

https://doi.org/10.3390/biom16020322

Pancreatic juice and bile were collected in the same manner as described for previous experiments. After sample collection for biochemical analyses, pancreatic juice was returned to the duodenum in all experimental conditions, whereas bile was returned only in control experiments.

Under conditions of eliminated bile flow, intraduodenal infusions of bile acid salt solutions were performed to assess concentration-dependent effects on pancreatic juice secretion. Infusions began with the onset of the feeding phase and continued until the end of the first hour of the post-feeding phase (Appendix A.2, Scheme A3).

Bile acid salt solutions were prepared using sodium salts of glycocholic acid, cholic acid, and deoxycholic acid derived from bovine bile (Sigma-Aldrich, St. Louis, MO, USA). Solutions with final concentrations of 0.02 M, 0.04 M, and 0.08 M were prepared by dissolving the appropriate amounts of bile acid salts in physiological saline and were administered intraduodenally at a constant rate of 7.5 mL/kg body mass/hour.

2.4. *Analysis of Pancreatic Juice and Bile*

2.4.1. Determination of Total Protein Content and Trypsin Activity in Pancreatic Juice

In all the studies performed, the volume of secreted pancreatic juice was measured, and the total protein content in it was determined by the Lowry et al. method, modified and adapted for analysis in microplates [22]. Bovine serum albumin (BSA—Bovine serum albumin, A-7638, Sigma Chemicals, Co., St. Louis, MO, USA) was used as the standard.

Trypsin activity was determined after activation of pancreatic juice with enterokinase (Sigma Chemicals, Co., USA) and using the BAPNA substrate (Na-benzoyl-DL-arginine-p-nitroanilide, Sigma) according to the method of Erlanger et al. adapted for microplate analysis [22,28]. The release of p-nitroaniline was monitored by measuring absorbance at a wavelength of 405 nm. Enzyme activity was expressed in units (U), defined as the amount of enzyme catalysing the hydrolysis of 1 µmol of substrate per minute [29].

2.4.2. Determination of Total Leptin Concentration in Bile

Leptin concentration in bile samples was determined by radioimmunoassay (RIA) using the Multi-Species Leptin RIA Kit (LINCO Research, Inc., St. Charles, MO, USA), which contains ^{125}I-labelled human leptin, primary antibody (anti-Multi-Species Leptin) from guinea pig serum, secondary antibody (anti-Guinea Pig IgG) from goat serum, 3% PEG, recombinant human leptin standards, and buffers. The antibodies are recommended for determining leptin in plasma of many species, including pigs (the lowest leptin concentration detected by the test is 1.0 ng/mL, and the antibodies show 67% affinity for porcine leptin). The radioactivity of standards and samples was analysed in a Packard–Cannbera gamma counter. Based on the readings of the standards, the counter automatically calculated the leptin concentration in the analysed samples.

2.5. *Determination of Cholecystokinin Concentration in Blood*

Blood samples of approximately 5 mL were collected into tubes containing K_3EDTA (tripotassium ethylenediaminetetraacetic acid, Sigma) and aprotinin (Traskolan, Jelfa, Jelenia Góra, Poland) 15 min before the feeding period and 15 min before the end of the feeding period. After centrifugation, plasma was collected and stored at −20 °C for CCK analysis. Cholecystokinin was determined in ethanol extracts using a ready-made radioimmunoassay kit (CCK RIA-Kit, Eurodiagnostica, Stockholm, Sweden) based on the double-antibody method. CCK-8s was used as the standard. The minimum sensitivity of the test was 0.3 pmol/L, and the intra-assay error was 6%. CCK was measured at pre-feeding and late-feeding time points.

https://doi.org/10.3390/biom16020322

2.6. Calculations and Statistical Analysis

Pancreatic juice secretion (volume, total protein concentration, and trypsin activity) and bile volume were expressed as output parameters and standardised to body mass (mL/kg body mass/h, mg/kg body mass/h, U/kg body mass/h, and mL/kg body mass/h, respectively). Pancreatic juice secretion during the perifeeding period, under conditions of either the presence or absence of pancreatic juice flow to the duodenum and following administration of pancreatin solutions with different trypsin activities (20%, 100%, and 200%), was expressed as the difference in secretion between the feeding and pre-feeding phases and between the post-feeding and pre-feeding phases. Data are presented as mean values $n = 6 \pm$ standard errors (SE).

After checking the normality of the data distribution using the Shapiro–Wilk test, statistical significance was analysed using the Friedman test (a nonparametric equivalent of one-way analysis of variance for repeated measures) and, if appropriate, the Wilcoxon test for dependent variables. Statistically significant differences were considered at a significance level of $p < 0.05$. Statistical analyses were performed using Statistica version 5.1 (StatSoft, Tulsa, OK, USA).

3. Results

Detailed protocol variants are summarised in Appendix A. For clarity, the Results are organised by regulatory question rather than by protocol number. Sections 3.1–3.3 report findings from Experimental Models I and II (Figure 2), respectively, focusing on evidence directly relevant to the primary feedback mechanisms.

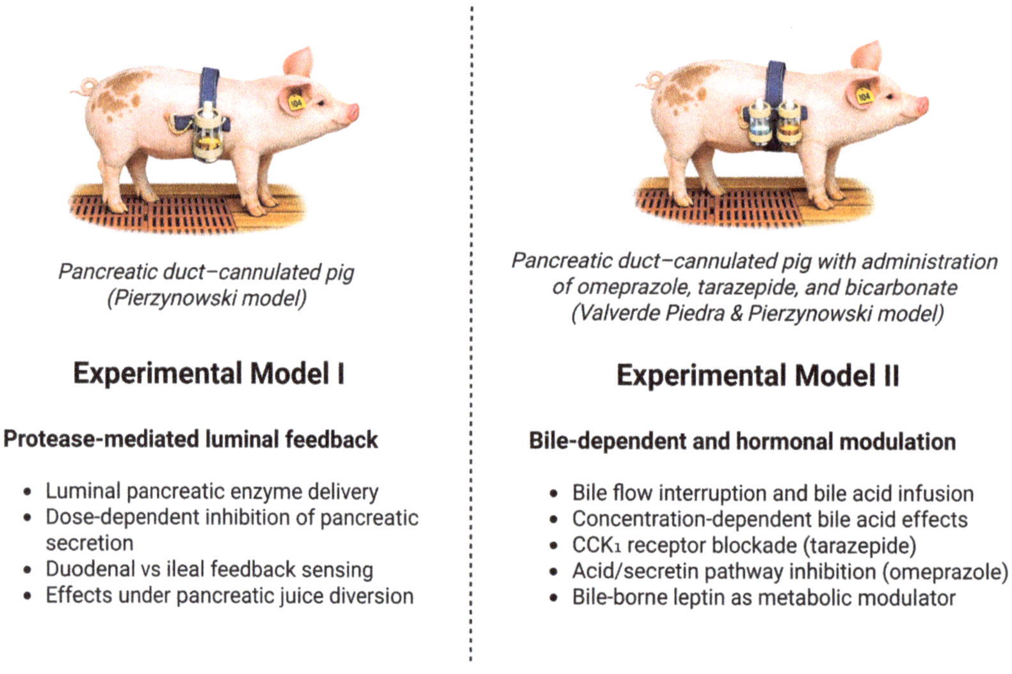

Figure 2. Overview of Experimental Models I and II and main experimental manipulations. Created using BioRender [30].

https://doi.org/10.3390/biom16020322

3.1. Effect of Pancreatic Enzymes in the Small Intestine on Pancreatic Juice Secretion

3.1.1. Pancreatic Juice Secretion Under Control and Interrupted Flow Conditions

Under control conditions with preserved pancreatic juice flow to the duodenum, basal pre-prandial pancreatic secretion was characterised by low pancreatic juice volume, protein output, and trypsin activity. Interruption of pancreatic juice delivery to the duodenum resulted in a significant increase in protein output and trypsin activity, while pancreatic juice volume showed only a numerical elevation (Table 2).

Table 2. Pancreatic juice (PJ) secretion and Cholecystokinin (CCK) concentration in blood plasma in piglets with preserved (PJ+) and eliminated (PJ−) pancreatic juice flow to the duodenum within pancreatic secretory phases (Values represent means $n = 6 \pm$ SE).

	Pancreatic Secretory Phase					
	Preprandial		Prandial		Postprandial	
	PJ+	PJ−	PJ+	PJ−	PJ+	PJ−
PJ output (mL/h/kg b.wt.)	1.71 ± 0.55	2.60 ± 0.16	3.20 ± 0.58 [a]	4.59 ± 0.50 [b]	2.76 ± 0.79 [a]	4.99 ± 0.41 [b]
PJ protein output (mg/h/kg b.wt.)	2.78 ± 0.49 [a]	4.87 ± 1.08 [b]	9.07 ± 1.82 [a]	28.2 ± 3.56 [b]	4.43 ± 1.17	12.7 ± 3.72
PJ trypsin output (U/h/kg b.wt.)	2.04 ± 0.41 [a]	4.87 ± 1.08 [b]	5.63 ± 0.73	15.4 ± 2.62	2.89 ± 0.61 [a]	7.69 ± 1.05 [b]
CCK (pmol/L)	1.36 ± 0.54	2.85 ± 2.11	1.79 ± 0.02	5.34 ± 3.94	-	-

[a,b]—different letters indicate statistically significant differences between preserved and interrupted PJ duodenal flow in the respective pancreatic secretory phases at the level of $p < 0.05$. (Wilcoxon test).

During the feeding phase, elimination of pancreatic juice flow led to a marked stimulation of pancreatic secretion, reflected by significantly higher pancreatic juice volume and protein output compared with preserved flow conditions. In the postprandial phase, pancreatic juice diversion maintained significantly elevated pancreatic juice volume and trypsin activity. Under preserved flow conditions, feed intake significantly stimulated protein and trypsin output, whereas pancreatic juice volume was not significantly affected (Table 2).

Plasma cholecystokinin (CCK) concentrations tended to be higher during interrupted pancreatic juice flow; however, due to high inter-individual variability, these differences did not reach statistical significance.

3.1.2. Effect of Intraduodenal Pancreatin Infusion Under Preserved Pancreatic Juice Flow

Under preserved pancreatic juice flow, feeding induced a physiological increase in pancreatic juice secretion. Intraduodenal administration of pancreatin solutions corresponding to 20%, 100%, and 200% of feeding-phase trypsin activity produced a dose-dependent tendency toward attenuation of the feeding-induced secretory response; however, these effects did not reach statistical significance during the feeding phase (Table 3).

In the post-feeding phase, pancreatin infusion at 100% and 200% trypsin activity resulted in significantly higher protein output and trypsin activity compared with control conditions, with values exceeding those observed during the feeding phase.

https://doi.org/10.3390/biom16020322

Table 3. Pancreatic juice (PJ) secretion during the feeding and post-feeding phase in piglets under conditions of preserved pancreatic juice flow and administration of pancreatin solutions to the duodenum (20%, 100%, 200% of trypsin activity). Changes in pancreatic juice secretion is presented as increments between feeding phase, post-feeding phase and pre-feeding phase values, respectively (means $n = 6 \pm$ SE).

	Pancreatin Intraduodenal Infusions Under Preserved PJ Flow Conditions			
Pancreatic secretory phase	100% id. PJ + id pancreatin infusions			
	0%	20%	100%	200%
	Pancreatic juice output (mL/h/kg b.wt.)			
Preprandial means	1.72 ± 0.12	0.98 ± 0.40	1.52 ± 0.27	1.44 ± 0.25
Prandial increments	1.48 ± 0.46	0.74 ± 0.52	0.20 ± 0.39	0.28 ± 0.39
Postprandial increments	1.05 ± 0.57	1.20 ± 0.63	1.02 ± 0.38	1.34 ± 0.72
	Pancreatic juice protein output increments (mg/h/kg b.wt.)			
Preprandial means	2.70 ± 0.40	6.90 ± 1.49	5.63 ± 0.96	4.50 ± 0.05
Prandial increments	6.29 ± 2.18	4.09 ± 1.89	2.85 ± 1.36	1.74 ± 0.32
Postprandial increments	1.65 ± 1.58 [a]	3.25 ± 0.92 [a]	9.44 ± 3.01 [b]	9.56 ± 4.00 [b]
	Pancreatic juice trypsin output increments (U/h/kg b.wt.)			
Preprandial means	2.04 ± 0.41	4.08 ± 0.54	3.86 ± 0.42	3.43 ± 0.12
Prandial increments	3.59 ± 0.98	2.04 ± 1.03	1.82 ± 0.83	1.39 ± 0.30
Postprandial increments	0.85 ± 0.81 [a]	1.97 ± 0.49 [a]	6.68 ± 2.61 [b]	7.53 ± 3.41 [b]

[a,b] = $p < 0.05$—different letters indicate statistically significant differences between respective pancreatic secretory phases of preserved and interrupted PJ duodenal flow (Wilcoxon test.).

3.1.3. Effect of Intraduodenal Pancreatin Infusion Under Interrupted Pancreatic Juice Flow

When pancreatic juice flow was interrupted, feeding produced a pronounced stimulation of pancreatic secretion. Intraduodenal pancreatin infusion significantly reduced this response in a concentration-dependent manner. Administration of pancreatin at 100% trypsin activity markedly suppressed protein output and trypsin activity compared with untreated animals (Table 4).

Table 4. Pancreatic juice (PJ) secretion during the feeding and post-feeding phase in piglets under conditions of interrupted pancreatic juice flow and administration of pancreatin solutions to the duodenum (20%, 100%, 200% of trypsin activity). Changes in pancreatic juice secretion are presented as increments between feeding phase, post-feeding phase and pre-feeding phase values, (means $n = 6 \pm$ SE).

	Pancreatin Intraduodenal Infusions Under Interrupted PJ Flow			
Pancreatic secretory phase	0% id. PJ + id pancreatin infusions			
	0%	20%	100%	200%
	Pancreatic juice output (mL/h/kg b.wt.)			
Preprandial mean	2.60 ± 0.08	1.90 ± 0.45	2.34 ± 0.59	1.43 ± 0.82
Prandial increments	1.99 ± 0.58	0.70 ± 0.41	0.26 ± 0.56	1.17 ± 0.74
Postprandial increments	1.93 ± 1.06	1.00 ± 0.54	1.01 ± 0.67	0.47 ± 0.75
	Pancreatic juice protein output increments (mg/h/kg b.wt.)			
Preprandial mean	11.6 ± 0.44	7.18 ± 2.42	10.1 ± 2.31	3.72 ± 1.31
Prandial increments	16.6 ± 4.00 [A]	4.38 ± 2.16 [AB]	1.46 ± 2.01 [B]	7.84 ± 4.74 [AB]
Postprandial increments	6.98 ± 2.02	3.06 ± 1.04	3.75 ± 1.45	3.20 ± 2.88
	Pancreatic juice trypsin output increments (U/h/kg b.wt.)			
Preprandial mean	4.80 ± 0.54	0.88 ± 0.36	3.74 ± 1.93	0.05 ± 0.3
Prandial increments	10.6 ± 2.99 [a]	3.92 ± 2.00 [ab]	1.07 ± 1.50 [b]	4.86 ± 3.25 [ab]
Postprandial increments	4.43 ± 1.72	2.62 ± 0.86	3.13 ± 1.01	2.48 ± 2.56

[a,b] = $p < 0.05$, [A,B] = $p < 0.01$. Different letters indicate statistically significant differences between treatments (Wilcoxon test).

https://doi.org/10.3390/biom16020322

Lower-dose pancreatin (20%) induced partial inhibition, whereas the highest dose (200%) showed reduced inhibitory efficacy. During the post-feeding phase, no significant differences among treatment groups were observed.

3.1.4. Effect of Ileal Pancreatin Infusion Under Preserved Pancreatic Juice Flow

Under preserved pancreatic juice flow, ileal infusion of pancreatin at all tested concentrations did not significantly alter pancreatic secretion during the feeding phase. In the post-feeding phase, a dose-dependent tendency toward increased protein output and trypsin activity was observed, reaching statistical significance only at the highest pancreatin concentration (200%) (Table 5).

Table 5. Pancreatic juice (PJ) secretion and cholecystokinin (CCK) concentration in peripheral blood plasma during the feeding and post-feeding phase in piglets under conditions of preserved pancreatic juice flow and administration of pancreatin solutions to the ileum (20%, 100%, 200% of trypsin activity). Changes in pancreatic juice secretion are presented as increments between feeding phase, post-feeding phase and pre-feeding phase values (means $n = 6 \pm$ SE).

Pancreatin Intraileal Infusions Under Preserved PJ Flow Conditions				
Pancreatic secretory phase	100% id. PJ + Ileal pancreatin infusion			
	0%	20%	100%	200%
Pancreatic juice output (mL/h/kg b.wt.)				
Preprandial means	1.72 ± 0.12	1.40 ± 0.25	1.45 ± 0.34	1.72 ± 0.09
Prandial increments	1.48 ± 0.46	0.32 ± 0.38	0.27 ± 0.43	1.21 ± 0.27
Postprandial increments	1.05 ± 0.57	0.20 ± 0.62	0.90 ± 0.34	1.05 ± 0.98
Pancreatic juice protein output increments (mg/h/kg b.wt.)				
Preprandial means	2.78 ± 0.40	10.8 ± 3.24	10.5 ± 3.41	6.88 ± 0.42
Prandial increments	6.29 ± 2.18	8.06 ± 3.64	7.77 ± 3.81	4.11 ± 1.00
Postprandial increments	1.65 ± 1.58 [a]	3.81 ± 1.45	5.23 ± 1.73	5.41 ± 2.48
Pancreatic juice trypsin output increments (U/h/kg b.wt.)				
Preprandial means	2.08 ± 0.41	7.04 ± 1.93	6.29 ± 2.28	4.56 ± 0.37
Prandial increments	3.59 ± 0.98	5.00 ± 2.33	4.25 ± 2.49	4.38 ± 2.26
Postprandial increments	0.85 ± 0.81 [a]	2.73 ± 1.10 [ab]	3.17 ± 1.25 [ab]	4.38 ± 2.26 [b]
Cholecystokinin (CCK) concentration in peripheral blood plasma (pmol/L)				
Prandial	1.36 ± 0.34	1.73 ± 0.31	2.62 ± 1.88	1.20 ± 0.27
Postprandial	1.79 ± 0.22	2.17 ± 0.81	2.60 ± 0.88	2.33 ± 0.80

[a,b] = $p < 0.05$. Different letters indicate statistically significant differences between treatments (Wilcoxon test).

Plasma CCK concentrations were not significantly affected by ileal pancreatin administration under these conditions.

3.1.5. Effect of Ileal Pancreatin Infusion Under Interrupted Pancreatic Juice Flow

Under conditions of interrupted pancreatic juice flow, ileal pancreatin infusion at all tested concentrations significantly reduced feeding-induced pancreatic enzyme secretion compared with untreated animals (Table 6). Protein output and trypsin activity were suppressed to levels comparable to those observed under preserved pancreatic juice flow.

The lowest pancreatin dose (20%) was sufficient to normalise pancreatic secretion. During the post-feeding phase, a significant reduction in protein output was observed at the 20% dose, whereas higher concentrations showed only numerical trends.

https://doi.org/10.3390/biom16020322

Table 6. Pancreatic juice (PJ) secretion during the feeding and post-feeding phase in piglets and cholecystokinin (CCK) concentration in peripheral blood plasma under conditions of interrupted pancreatic juice flow and administration of pancreatin solutions to the ileum (20%, 100%, 200% of trypsin activity). Pancreatic juice secretion is presented as increments between feeding phase, post-feeding phase and pre-feeding phase values (means $n = 6 \pm$ SE).

	Pancreatin Intraileal Infusions Under Interrupted PJ Flow			
	0% id PJ + ileal pancreatin infusion			
Pancreatic secretory phase	0%	20%	100%	200%
	Pancreatic juice output (mL/h/kg b.wt.)			
Preprandial means	2.60 ± 0.08	1.73 ± 0.33	1.51 ± 0.63	2.57 ± 0.85
Prandial increments	1.99 ± 0.58	0.87 ± 0.27	1.08 ± 0.55	0.03 ± 0.78
Postprandial increments	1.93 ± 1.06	1.24 ± 0.38	1.13 ± 0.53	0.89 ± 0.84
	Pancreatic juice protein output increments (mg/h/kg b.wt.)			
Preprandial means	11.6 ± 0.44	7.93 ± 2.09	6.84 ± 1.31	6.64 ± 1.75
Prandial increments	16.6 ± 4.00 ^{A}	3.63 ± 1.30 ^{b}	4.73 ± 2.87 ^{b}	4.92 ± 3.07 ^{b}
Postprandial increments	6.98 ± 2.02	2.02 ± 0.57 ^{b}	6.38 ± 4.12 ^{ab}	3.52 ± 3.40 ^{ab}
	Pancreatic juice trypsin output increments (U/h/kg b.wt.)			
Preprandial means	4.80 ± 0.54	3.23 ± 1.17	3.18 ± 0.80	2.81 ± 0.45
Prandial increments	10.6 ± 2.99 ^{a}	1.58 ± 0.62 ^{b}	1.63 ± 0.98 ^{b}	1.99 ± 1.74 ^{b}
Postprandial increments	4.43 ± 1.72	1.16 ± 0.32	3.38 ± 2.13	2.92 ± 2.08
	Cholecystokinin (CCK) concentration in peripheral blood plasma (pmol/L)			
Prandial	2.85 ± 2.11	1.61 ± 0.47	2.90 ± 1.59	4.33 ± 0.57
Postprandial	5.34 ± 3.94	2.51 ± 0.78	3.39 ± 1.21	3.21 ± 0.78

^{a,b} = $p < 0.05$, ^{A} = $p < 0.01$. Different letters indicate statistically significant differences between treatments (Wilcoxon test).

3.1.6. Cholecystokinin Concentration in Peripheral Blood Plasma

Across all experimental conditions, plasma CCK concentrations did not differ significantly between preserved and interrupted pancreatic juice flow. Although higher mean CCK values were observed during interrupted flow, particularly in the feeding phase, substantial inter-individual variability prevented statistical confirmation. Neither duodenal nor ileal pancreatin infusion produced significant alterations in circulating CCK levels (Tables 5 and 6).

3.2. *Effect of Bile, Bicarbonate, and Pharmacological Interventions on Pancreatic and Bile Secretion*

3.2.1. Pancreatic Juice and Bile Secretion with Bile Flow Interruption and Bicarbonate Infusion

Under control conditions with preserved pancreatic juice and bile flow, pre-feeding pancreatic juice volume was 1.62 ± 0.20 mL/kg body mass/hour, protein output was 4.64 ± 1.33 mg/kg body mass/hour, and trypsin activity reached 2.44 ± 0.68 U/kg body mass/hour.

Interruption of bile flow to the duodenum significantly increased pancreatic juice volume to 3.00 ± 0.18 mL/kg body mass/hour and trypsin activity to 4.16 ± 0.28 U/kg body mass/hour ($p < 0.05$). In contrast, pancreatic protein output remained comparable to control values. Combined interruption of both bile and pancreatic juice flow produced a similar stimulatory effect on pancreatic secretion (Table 7).

During the feeding phase, bile flow interruption resulted in significantly higher pancreatic protein output (18.8 ± 2.13 mg/kg versus 12.9 ± 1.96 mg/kg in controls, $p < 0.05$) and increased trypsin activity (11.2 ± 1.14 U/kg versus 7.67 ± 1.49 U/kg, $p < 0.05$). Intraduodenal bicarbonate infusion under conditions of bile diversion showed a tendency toward further stimulation of pancreatic secretion, although these changes did not consistently reach statistical significance.

https://doi.org/10.3390/biom16020322

Table 7. The Influence of interruption of bile and pancreatic juice duodenal flow and the effect of omeprazole iv. (OMP) and tarazepide id. (TA) administration on pancreatic juice secretion and bile outflow ($n = 6 \pm$ SEM).

Pancreatic Secretory Phase	Treatment					
	Bile+, PJ+ (Control)	**Bile−, Pj+**	**Bile−, Pj−**	**Bile+, PJ+, OMP+**	**Bile+, PJ+, TA+**	**Bile+, PJ+, OMP+, TA+**
	Pancreatic juice output (mL/h/kg b.wt.)					
Preprandial	1.6 ± 0.3	2.8 ± 0.1 *	3.3 ± 0.8 *	2.5 ± 0.7	2.2 ± 0.5	1.3 ± 0.5
Prandial	3.3 ± 0.6	4.9 ± 0.9	4.8 ± 0.7	2.2 ± 0.6	2.9 ± 0.7	2.0 ± 0.8
Postprandial	2.4 ± 0.6	4.2 ± 0.7 *	4.1 ± 0.8 *	0.9 ± 0.1 *	2.0 ± 0.6	0.8 ± 0.1 *
	Pancreatic juice protein output (mg/h/kg b.wt.)					
Preprandial	2.1 ± 0.5	6.7 ± 2.5	8.3 ± 3.0 *	4.2 ± 1.4	2.6 ± 1.4	2.6 ± 0.9
Prandial	9.7 ± 1.9	15.7 ± 5.6	16.0 ± 4.6	9.2 ± 3.7	5.4 ± 3.7 *	7.5 ± 1.9
Postprandial	4.3 ± 1.1	8.5 ± 3.0	9.7 ± 2.3 *	5.9 ± 1.7	4.1 ± 1.7	7.1 ± 1.6
	Pancreatic juice trypsin output (U/h/kg b.wt.)					
Preprandial	1.1 ± 0.3	4.8 ± 1.7	4.6 ± 1.8 *	2.4 ± 1.1	1.3 ± 0.4	0.9 ± 0.7
Prandial	5.3 ± 1.1	11.0 ± 4.2	11.1 ± 3.7	5.3 ± 2.7	2.9 ± 1.1	3.4 ± 1.4
Postprandial	2.6 ± 0.7	6.7 ± 2.5	7.3 ± 2.2 *	3.1 ± 1.4	2.0 ± 0.7	2.8 ±1.5
	Bile outflow (mL/h/kg b.wt.)					
Preprandial	2.9 ± 0.7	4.3 ± 0.7	4.4 ± 0.5 *	3.8 ± 0.3	4.1 ± 0.7	2.7 ± 0.4
Prandial	3.8 ± 0.7	3.9 ± 0.6	3.8 ± 0.3	3.0 ± 0.4	4.0 ± 0.7	2.9 ± 0.6
Postprandial	4.1 ± 0.6	3.2 ± 0.3	3.1 ± 0.5	2.6 ± 0.4 *	3.2 ± 0.6	2.5 ± 0.2 *

*—$p < 0.05$ differences between experimental and control treatments at the same secretory phase (Wilcoxon test).

In the post-feeding phase, interruption of bile flow alone or combined bile and pancreatic juice diversion maintained elevated pancreatic secretion. The highest protein output was observed under combined interruption conditions (17.1 ± 2.48 mg/kg versus 8.03 ± 2.50 mg/kg in controls, $p < 0.05$) (Table 7).

3.2.2. Effect of Gastric Acid Inhibition and CCK1 Receptor Blockade

Intravenous administration of omeprazole (1 mg/kg body mass) 30 min before feeding did not significantly affect pre-feeding pancreatic secretion. However, a tendency toward reduced trypsin activity and increased pancreatic juice volume was observed.

Intraduodenal administration of tarazepide (5 mg/kg body mass), a selective CCK_1 receptor antagonist, tended to reduce pancreatic protein output and trypsin activity, while bile secretion showed a tendency to increase. Combined administration of omeprazole and tarazepide resulted in higher pancreatic juice volume accompanied by lower protein output and trypsin activity during the pre-feeding phase (Table 7).

During the feeding phase, omeprazole reduced pancreatic juice volume to 2.51 ± 0.43 mL/kg compared with 3.82 ± 0.43 mL/kg in control animals. Tarazepide caused a more pronounced reduction in pancreatic juice volume (2.78 ± 0.55 mL/kg) and was associated with a marked decrease in protein output (5.89 ± 0.93 mg/kg versus 12.9 ± 1.96 mg/kg in controls). Combined treatment further reduced pancreatic juice volume as well as protein and trypsin output. A significant reduction in bile flow was observed only following combined omeprazole and tarazepide administration (2.99 ± 0.36 mL/kg versus 3.96 ± 0.54 mL/kg in controls, $p < 0.05$).

In the post-feeding phase, all pharmacological treatments resulted in lower pancreatic secretion compared with control animals. The lowest pancreatic juice volume was observed following combined omeprazole and tarazepide treatment (0.92 ± 0.17 mL/kg). Under these conditions, bile flow was also significantly reduced (2.23 ± 0.20 mL/kg versus 4.05 ± 0.70 mL/kg in controls, $p < 0.05$) (Table 7).

https://doi.org/10.3390/biom16020322

3.3. *Effect of Bile Flow Interruption and Bile Acid Salt Infusion on Pancreatic and Bile Secretion*

3.3.1. Effect of Prolonged Bile Flow Interruption

In experiments with extended observation, control pancreatic juice secretion during the pre-feeding phase averaged 1.54 ± 0.20 mL/kg body mass/hour, with protein output of 3.41 ± 0.36 mg/kg body mass/hour.

Interruption of bile flow for 10 h significantly increased pancreatic juice volume to 3.29 ± 0.37 mL/kg body mass/hour and protein output to 5.85 ± 0.85 mg/kg body mass/hour ($p < 0.05$). At the same time, bile secretion was significantly reduced under interrupted conditions (1.94 ± 0.27 mL/kg body mass/hour versus 2.36 ± 0.26 mL/kg in controls) (Table 8).

Table 8. Pancreatic juice secretion of control (connected bile and PJ tubbing), bicarbonates infused instead of bile during the exp. and after overnight (13 h) bile diversion and bile or bile salts id. administration at various concentrations within 2 h (Mean values from n = 6 ± SEM).

Pancreatic Secretory Phase	Treatment						
	Control	B−, PJ+	B−, PJ+, ($NaHCO_3^-$)+	B−, PJ+, BAS (0.02 M)	B−, PJ+, BAS (0.04 M)	B−, PJ+, BAS (0.08 M)	B−, PJ+, B+
	Pancreatic juice outflow (mL/h/kg b.wt.)						
Preprandial	1.0 ± 0.4	2.9 ± 0.8	3.5 ± 0.7 *	2.7 ± 0.5 *	3.3 ± 0.6	3.3 ± 0.7	1.5 ± 0.4
Prandial	3.5 ± 0.3	4.8 ± 0.6	4.8 ± 0.7	4.1 ± 0.6	3.2 ± 0.5	2.8 ± 1.0	3.5 ± 0.6
Postprandial 1st h	2.1 ± 0.1	4.6 ± 0.6 **	4.2 ± 0.6 *	2.3 ± 0.1	3.0 ± 0.5	2.4 ± 0.9	3.6 ± 0.8
Postprandial 2nd h	2.0 ± 0.3	4.3 ± 0.6 **	4.8 ± 0.5 ***	3.4 ± 0.5 *	2.6 ± 0.5	3.7 ± 1.3	3.2 ± 0.5 *
Postprandial 3rd h	2.4 ± 0.4	3.7 ± 0.5	6.6 ± 0.4 ***	3.8 ± 0.5 *	3.3 ± 0.4	3.6 ± 0.5	2.9 ± 0.6
	Pancreatic juice protein outflow (mg/h/kg b.wt.)						
Preprandial	3.0 ± 1.0	6.9 ± 1.8	7.2 ± 1.0 *	4.8 ±1.5	5.6 ± 1.6	9.3 ± 1.5 ***	4.3 ± 1.0
Prandial	36.6 ± 15.1	52.7 ± 12.8	20.8 ± 7.4	24.1 ± 8.1	23.1 ± 7.1	11.4 ± 5.1	25.6 ± 6.5
Postprandial 1st h	11.1 ± 2.0	28.2 ± 5.5 *	13.3 ± 2.8	14.3 ± 1.37	11.9 ± 2.9	9.3 ± 3.4	22.7 ± 6.8
Postprandial 2nd h	10.4 ± 2.7	23.8 ± 4.3 *	10.5 ± 0.8	14.9 ± 1.73	13.1 ± 3.1	11.5 ± 5.2	20.6 ± 5.4
Postprandial 3rd h	11.9 ± 3.1	26.7 ± 4.3 **	13.6 ± 1.5	15.2 ± 2.06	13.1 ± 3.3	11.2 ± 2.3	21.0 ±5.9
	Pancreatic juice trypsin outflow (U/h/kg b.wt.)						
Preprandial	2.2 ± 0.9	45.5 ± 1.4	4.5 ± 1.5	3.7 ± 2.1	7.3 ± 2.1 *	6.7 ± 1.4 *	2.9 ± 0.8
Prandial	12.7 ± 3.0	39.7 ± 10.2 *	14.2 ± 6.8	16.5 ± 5.8	12.3 ± 4.4	9.1 ± 4.3	22.3 ±8.3
Postprandial 1st h	9.0 ± 2.0	23.0 ± 9.1 *	10.3 ± 4.0	9.7 ± 3.3	8.7 ± 3.0	7.8 ± 3.5	17.7 ± 5.6
Postprandial 2nd h	8.1 ± 2.3	18.7 ± 3.6 *	6.4 ± 1.9	11.4 ± 4.0	8.8 ± 2.7	9.3 ±5.0	16.1 ± 4.5
Postprandial 3rd h	8.5 ± 2.6	19.3 ± 2.5 *	7.6 ± 2.4	10.5 ± 3.6	9.6 ± 3.0	9.1 ± 2.7	13.9 ± 4.4
	Bile outflow (mL/h/kg b.wt.)						
Preprandial	2.5 ± 0.3	2.1 ± 0.4	5.0 ± 2.0	3.6 ± 0.8	3.1 ± 0.5 *	2.5 ± 0.7	1.6 ± 0.1 *
Prandial	3.7 ± 0.4	1.4 ± 0.1 **	5.0 ± 1.2	2.3 ± 0.3 *	3.1 ± 0.3 *	2.5 ± 0.7	1.9 ± 0.1 *
Postprandial 1st h	3.7 ± 0.5	1.9 ± 0.3 *	3.7 ± 1.2	2.6 ± 0.3 *	4.0 ± 0.5	3.8 ± 0.7	3.3 ± 0.4
Postprandial 2nd h	3.0 ± 0.6	2.2 ± 0.4	3.3 ± 0.7	3.2 ± 0.4	3.1 ± 0.3	4.5 ± 1.3	3.7 ± 0.3
Postprandial 3rd h	3.3 ± 0.5	1.6 ± 0.3 *	4.1 ± 1.0	2.9 ± 0.3	3.9 ± 0.8	5.0 ± 0.6	2.5 ± 0.3

* = $p < 0.05$, ** = $p < 0.01$, *** = $p < 0.001$ differences between treatments at the same secretory phase (Wilcoxon test).

During the feeding phase, pancreatic juice volume increased from 3.83 ± 0.45 mL/kg body mass/hour under control conditions to 4.89 ± 0.34 mL/kg body mass/hour during bile diversion. Although this increase did not reach statistical significance, protein output and trypsin activity were markedly elevated (32.0 ± 7.12 mg/kg versus 14.6 ± 1.67 mg/kg and 24.5 ± 7.79 U/kg versus 6.97 ± 0.97 U/kg, respectively; both $p < 0.05$). In parallel, bile secretion was strongly suppressed to 1.17 ± 0.17 mL/kg body mass/hour ($p < 0.05$).

In the first hour following feeding, pancreatic juice volume remained numerically higher under bile flow interruption (4.29 ± 0.45 mL/kg versus 2.43 ± 0.36 mL/kg), while protein output and trypsin activity were significantly increased compared with control animals (24.7 ± 5.07 mg/kg versus 11.3 ± 1.74 mg/kg and 16.8 ± 4.86 U/kg versus 5.33 ± 0.62 U/kg, respectively; $p < 0.05$). Bile secretion remained significantly reduced (1.76 ± 0.23 mL/kg versus 3.67 ± 0.40 mL/kg, $p < 0.05$).

During the second and third post-feeding hours, pancreatic juice volume remained elevated under bile diversion (3.87 ± 0.38 and 3.34 ± 0.42 mL/kg body mass/hour, respectively). Protein output remained significantly higher throughout this period

https://doi.org/10.3390/biom16020322

(20.6 ± 4.25 mg/kg in hour two and 22.9 ± 3.81 mg/kg in hour three, $p < 0.05$). Trypsin activity showed sustained elevation, although statistical significance was not reached at time points. Bile secretion remained significantly suppressed across all post-feeding hours (Table 8).

3.3.2. Effect of Intraduodenal Bile Acid Salt Infusion

Intraduodenal infusion of bile acid salt solutions at concentrations of 0.02 M, 0.04 M, and 0.08 M under conditions of interrupted bile flow modulated pancreatic secretion in a concentration-dependent manner.

During the feeding phase, infusion of 0.02 M and 0.04 M bile acid solutions resulted in significantly increased pancreatic protein output and trypsin activity compared with pre-feeding values ($p < 0.05$). In contrast, administration of the highest concentration (0.08 M) failed to stimulate pancreatic secretion in response to food intake (Table 8).

In the post-feeding phase, pancreatic protein output and trypsin activity displayed clear dose-dependent patterns. Animals receiving the 0.02 M solution maintained significantly elevated secretion throughout the 3 h observation period ($p < 0.05$). In contrast, infusion of 0.04 M bile acids resulted in partial attenuation of postprandial secretion, with values approaching pre-feeding levels. The highest bile acid concentration (0.08 M) produced suppression of the postprandial pancreatic secretory response.

Bile volume did not differ significantly among treatment groups during the feeding phase. However, during the post-feeding period, bile outflow increased significantly, with the highest concentrations producing the greatest elevations. The 0.08 M solution induced the most pronounced postprandial increase in bile secretion, reaching statistical significance during post-feeding hours one to three (Table 8).

3.4. Leptin Concentration in Bile

Leptin concentration in bile under control conditions, with preserved pancreatic juice and bile flow to the duodenum, ranged between 16.5 and 19.0 ng/mL and did not differ significantly between interdigestive and digestive periods. Interruption of bile flow resulted in lower biliary leptin concentrations, ranging from 11.7 to 14.5 ng/mL.

During the interdigestive period, leptin secretion in control animals averaged 40.6 ± 12.2 ng/kg body mass/hour. During the feeding phase, leptin secretion increased to 75.4 ± 24.7 ng/kg body mass/hour, although this increase did not reach statistical significance. Elevated leptin secretion was maintained during the first and second post-feeding hours (69.5 ± 17.5 and 62.4 ± 5.5 ng/kg body mass/hour, respectively).

Under conditions of interrupted bile flow, leptin secretion during the feeding phase was significantly reduced to 15.3 ± 5.7 ng/kg body mass/hour ($p < 0.05$), corresponding to an approximately 50% decrease compared with control values. In the first and second hours following feeding, leptin secretion remained significantly lower under bile diversion (23.0 ± 10.4 and 20.0 ± 7.4 ng/kg body mass/hour, respectively; $p < 0.05$). A gradual recovery toward baseline values was observed during the third post-feeding hour (Table 9).

Table 9. Leptin secretion in bile (ng/kg body mass/hour) under conditions of preserved and interrupted bile flow to the duodenum in piglets (Mean values from $n = 6 \pm$ SE).

Items	Piglets with Preserved PJ and Bile Flow to Duodenum	Piglets with Interrupted Bile Flow to Duodenum
Before feed intake	40.6 ± 12.2	33.0 ± 6.6
Feeding phase	75.4 ± 24.7 ^{a}	15.3 ± 5.7 ^{*b}
Post-feeding phase (1 h)	69.5 ± 17.5 ^{a}	23.0 ± 10.4 ^{b}
Post-feeding phase (2 h)	62.4 ± 5.5 ^{a}	20.0 ± 7.4 ^{b}
Post-feeding phase (3 h)	58.0 ± 21.9	19.6 ± 8.7

^{a,b} = $p < 0.05$. Different letters indicate statistically significant differences between treatments (Wilcoxon test).
* = $p < 0.05$ indicates significant difference from the pre-feeding sample.

https://doi.org/10.3390/biom16020322

4. Discussion

The present study evaluates several feedback mechanisms involved in the regulation of pancreatic juice secretion in piglets and provides integrated physiological observations on the interactions between pancreatic enzyme-dependent feedback, bile acid-related signalling, hormonal regulation, and metabolic influences [1,2,31]. These observations extend existing concepts of exocrine pancreatic regulation, which have been derived largely from rodent models [1,32,33], by examining these mechanisms in a large-animal model that displays closer physiological similarity to humans, particularly with regard to migrating motor complex (MMC) periodicity and patterns of pancreatic enzyme secretion [34–36].

Taken together, the results support the interpretation that pancreatic exocrine secretion is governed by multiple interacting regulatory pathways. These include protease-dependent luminal feedback involving trypsin-mediated inactivation of CCK-releasing peptides, as well as modulatory influences related to bile acids, hormone-dependent signalling (CCK and secretin), and metabolic factors such as leptin. Rather than acting as isolated mechanisms, these pathways appear to function in a coordinated and partially compensatory manner, allowing pancreatic secretion to adapt to alterations in individual regulatory inputs. Given the breadth of experimental variants, the discussion is intentionally focused on findings that directly inform the central feedback mechanisms governing pancreatic exocrine regulation.

4.1. The Importance of Pancreatic Juice in Feedback Regulation of the Exocrine Pancreas

Interruption of pancreatic juice flow to the duodenum was associated with a marked increase in pancreatic enzyme secretion, reflected by elevated protein output and trypsin activity, particularly during the preprandial phase. This response is consistent with the classical concept of protease-dependent negative feedback, whereby luminal trypsin limits pancreatic secretion through proteolytic inactivation of cholecystokinin (CCK)-releasing factor (LCRF) and monitor peptide (MP) [4,6,37,38]. Under physiological conditions, the return of pancreatic juice to the duodenum promotes degradation of these peptides, thereby reducing stimulation of intestinal I cells and limiting further enzyme release [1,37–39]. Disruption of this mechanism permits accumulation of CCK-releasing peptides and consequently enhances pancreatic secretion.

Despite the pronounced stimulation of enzyme output following pancreatic juice diversion, circulating plasma CCK concentrations exhibited high inter-individual variability and did not reach statistical significance under several experimental conditions. This dissociation between peripheral CCK levels and pancreatic secretory responses highlights a limitation inherent to endocrine measurements in large-animal models. Given the short plasma half-life of CCK and the predominant role of paracrine and vagally mediated signalling, circulating CCK concentrations may not accurately reflect functional activation of CCK-dependent regulatory pathways at the intestinal–neural interface [5,6,40].

Further support for a protease-dependent feedback mechanism was provided by the dose-dependent inhibition observed during intraduodenal pancreatin administration. Maximal suppression of pancreatic protein output occurred at pancreatin doses corresponding to physiological trypsin activity, whereas supraphysiological concentrations were associated with partial recovery of secretion. This biphasic response suggests saturable feedback kinetics and may reflect depletion of luminal CCK-releasing peptides, receptor desensitisation, or recruitment of compensatory regulatory pathways. Such behaviour is consistent with current concepts that protease-dependent feedback constitutes one component of a broader, non-linear regulatory network rather than a simple on–off mechanism [1,2,31].

A notable finding of the present study is the effectiveness of protease-dependent feedback at distal intestinal sites. Ileal delivery of pancreatin under conditions of interrupted pancreatic juice flow significantly reduced pancreatic enzyme secretion to levels comparable with those observed during preserved duodenal return. This observation challenges the traditional duodenum-centred view of pancreatic feedback regulation and supports the concept of distributed enteroendocrine sensing along the length of the small intestine [5,34,39]. The sensitivity of ileal feedback, particularly at relatively low pancreatin doses, suggests that protease-dependent regulation is not confined to proximal intestinal segments.

These findings are compatible with an extension of the ileal brake concept beyond its established role in gastrointestinal motility and nutrient absorption to include modulation of pancreatic exocrine secretion. From a translational perspective, this mechanism may be relevant in clinical settings such as pancreatic enzyme replacement therapy, where distal delivery of active enzymes could contribute to suppression of endogenous pancreatic secretion [5,39].

Pharmacological inhibition of CCK_1 receptors provided additional evidence for the involvement of CCK-dependent pathways in pancreatic enzyme regulation. Administration of tarazepide resulted in a substantial reduction in prandial enzyme output, indicating that a significant proportion of food-stimulated pancreatic secretion in piglets depends on CCK_1-mediated signalling. Given the reported predominance of CCK_2 receptors on porcine acinar cells, this effect is most likely mediated indirectly via CCK_1 receptors located on vagal afferent fibres, consistent with the established role of neural amplification mechanisms in large mammals [1,19,41–43].

In contrast, inhibition of gastric acid secretion with omeprazole primarily reduced pancreatic juice volume without significantly affecting enzyme output. This pattern supports the established role of acid-stimulated secretin release in regulating ductal bicarbonate secretion while indicating that pancreatic enzyme output is governed predominantly by CCK-dependent and neural mechanisms rather than by acid–secretin signalling alone [39,44].

The most pronounced inhibitory effect on pancreatic enzyme secretion was observed following combined blockade of gastric acid secretion and CCK_1 receptors. The additive nature of this suppression indicates that pancreatic exocrine secretion is controlled by parallel regulatory pathways rather than by a single dominant mechanism [13,16,38]. The concurrent reduction in bile output further suggests coordinated regulation of pancreatic enzyme secretion and gallbladder emptying, ensuring synchronised delivery of digestive enzymes and bile salts during nutrient digestion.

4.2. The Importance of Bile in Feedback Regulation of the Exocrine Pancreas

One of the most prominent observations of the present study was the sustained stimulation of pancreatic enzyme secretion following interruption of bile flow to the duodenum. Under these conditions, both pancreatic protein output and trypsin activity increased markedly during the prandial phase and remained elevated throughout the postprandial period. This prolonged hypersecretion indicates that bile-derived signals exert an inhibitory influence on pancreatic exocrine secretion under physiological conditions.

Previous experimental studies have suggested an inhibitory role of bile acids in the regulation of pancreatic enzyme secretion; however, the magnitude and physiological relevance of this effect have varied considerably between species and experimental models. The present findings are consistent with the concept that physiological exposure to bile acids provides a tonic suppressive signal limiting excessive pancreatic enzyme secretion, whereas removal of this signal leads to compensatory hypersecretion. In this context, bile acids may contribute to feedback regulation by adjusting pancreatic output once lipid digestion has been initiated [13,42,45].

Infusion of bile acid salts at graded concentrations revealed a clear concentration-dependent modulation of pancreatic secretion. Lower concentrations (0.02 M) maintained

https://doi.org/10.3390/biom16020322

or modestly enhanced enzyme output, whereas higher concentrations (0.08 M) markedly suppressed postprandial pancreatic secretion. Intermediate concentrations (0.04 M) produced transitional responses between these two extremes. This biphasic pattern indicates that bile acids do not function as simple inhibitory signals but rather modulate pancreatic secretion in a dose-dependent manner [44,46,47].

Several bile acid–responsive signalling pathways have been proposed to mediate such effects, including the farnesoid X receptor (FXR/NR1H4) and the Takeda G-protein–coupled bile acid receptor (TGR5/GPBAR1). Activation of FXR in intestinal epithelial cells has been linked to induction of fibroblast growth factor 15/19 (FGF15 in rodents, FGF19 in humans), whereas TGR5 activation has been associated with modulation of enteroendocrine signalling. Importantly, neither FXR nor TGR5 signalling was directly assessed in the present study, and their involvement should therefore be regarded as hypothetical. The present results demonstrate a physiological phenomenon rather than a defined molecular mechanism [13,35,45,48].

The bile acid concentrations applied in this study (20–80 mM) approximate luminal levels encountered in the proximal small intestine following postprandial gallbladder emptying and exceed systemic portal concentrations by one to two orders of magnitude. This spatial gradient is physiologically relevant, as intestinal epithelial cells, enteroendocrine cells, and pancreatic ductal epithelium are capable of sensing luminal bile acid exposure independently of systemic bile acid signalling [11,46].

Collectively, these findings demonstrate that bile acids modulate pancreatic exocrine secretion in a concentration-dependent manner in vivo. While the underlying receptor mechanisms remain unresolved, the results provide a functional framework supporting bile acids as modulatory signals within the multi-level feedback system governing pancreatic secretion [14,35,45,48]. Future studies employing receptor-selective pharmacological tools and direct measurement of downstream signalling pathways will be required to delineate the relative contributions of FXR-, TGR5-, and alternative mechanisms.

4.3. Leptin in Bile

The present study provides novel observational data on leptin concentrations in bile under different experimental conditions. To our knowledge, biliary leptin secretion has not previously been characterised in a chronic large-animal model. The results demonstrate a temporal association between changes in biliary leptin output and alterations in pancreatic exocrine secretion; however, no causal relationship can be inferred from the present data [19,49,50].

Leptin is primarily recognised as a metabolic hormone involved in central regulation of energy balance and glucose homeostasis. Experimental studies have demonstrated that leptin-responsive neuronal pathways, including projections to the dorsal motor nucleus of the vagus, may link metabolic status with pancreatic endocrine and exocrine function. This provides a conceptual framework for indirect modulation of pancreatic activity by metabolic signals, although such mechanisms were not directly addressed in the present study. In vitro and ex vivo studies have suggested that leptin may exert concentration-dependent effects on pancreatic enzyme secretion. Physiological leptin concentrations have been reported to potentiate CCK-stimulated acinar responses, whereas supraphysiological levels may inhibit enzyme secretion through CCK1 receptor-dependent vagal mechanisms. Importantly, these observations originate primarily from isolated tissue and cell culture models and cannot be directly extrapolated to the in vivo conditions applied here [6,9,18,19,42,43].

In the current study, interruption of bile flow resulted in an approximately 50% reduction in biliary leptin output, concomitant with a marked increase in pancreatic enzyme secretion.

https://doi.org/10.3390/biom16020322

Although this inverse relationship is intriguing, it remains purely correlative. Several alternative mechanisms may account for pancreatic hypersecretion during bile diversion, including loss of bile acid–dependent inhibitory signalling, altered intestinal luminal conditions, changes in CCK release kinetics, and simultaneous disruption of multiple feedback pathways [13,36,42,45].

Accordingly, the present findings should be regarded as hypothesis-generating rather than mechanistically conclusive. Biliary leptin may represent a metabolic signal associated with bile-dependent regulation of digestive processes rather than a primary regulator of pancreatic exocrine secretion. Future studies employing direct functional manipulation of leptin signalling, combined with assessment of bile acid receptor pathways, will be required to determine whether leptin contributes to phase-dependent fine-tuning of pancreatic exocrine function.

The experimental findings are consistent with a hierarchically organised framework of pancreatic feedback regulation, in which protease-dependent luminal feedback represents the dominant inhibitory component, while bile- and metabolism-related pathways exert modulatory influences on pancreatic exocrine secretion (Table 10).

Table 10. Summary of multilevel feedback regulation of pancreatic exocrine secretion in piglets.

Level	Signal	Core Observation	Interpretation
I	Pancreatic proteases	PJ diversion ↑ secretion 2–3×; pancreatin restores inhibition	Primary inhibitory feedback
II	Distal protease sensing	Ileal pancreatin suppresses secretion	Feedback beyond duodenum
III	CCK/secretin pathways	Tarazepide ↓ ~54%; combined blockade ↓ ~65%	Parallel effector pathways
IV	Bile acids/leptin	Bile diversion ↑ secretion; bile acids modulate; leptin ↓	Phase-dependent modulation

↑ indicates an increase in pancreatic exocrine secretion, whereas ↓ indicates a decrease or inhibition of secretion.

4.4. Species Specific Considerations: Pigs Versus Rodents and Humans

The pig represents an intermediate translational model between rodents and humans with respect to pancreatic exocrine physiology. This is particularly evident in the periodicity of the migrating motor complex (MMC) and its temporal coordination with pancreatic secretion. In contrast to rodents, which exhibit short MMC cycles (~20 min), and humans, in whom cycles typically last ~90 min, pigs display an intermediate periodicity of approximately 60 min. This temporal organisation allows investigation of regulatory mechanisms operating over longer physiological time scales, which may not be adequately captured in rodent models [44,51,52]. Importantly, several key features of pancreatic exocrine regulation in pigs, including meal size, bile acid composition, and the organisation of postprandial secretory phases, more closely resemble human physiology than rodent models, supporting the translational relevance of porcine findings.

The robust negative feedback responses observed in the present study support the physiological relevance of protease-dependent regulation in large mammals. Moreover, the finding that both duodenal and ileal administration of pancreatin effectively suppressed pancreatic enzyme secretion suggests that feedback sensing is not restricted to the proximal intestine but may operate along an extended segment of the small intestine. This observation is consistent with the concept of distributed enteroendocrine sensing of luminal proteolytic activity and may contribute to species-specific differences reported in the literature [16,34,44,52]. Such distributed sensing mechanisms may be particularly relevant to human conditions characterised by altered intraluminal protease activity, including pancreatic exocrine insufficiency, enzyme replacement therapy, and post-surgical alterations of intestinal transit.

Despite these translational advantages, caution is required when extrapolating pharmacological findings to humans. Species-specific differences in drug absorption, hepatic

https://doi.org/10.3390/biom16020322

metabolism, receptor expression, and neural organisation may substantially influence the magnitude and mechanisms of pancreatic responses. In particular, pigs have been reported to exhibit predominant CCK-2 receptor expression on pancreatic acinar cells, whereas humans display higher CCK-1 receptor expression on both acinar cells and vagal afferent terminals [9,15,23,53,54].

In the present study, pharmacological blockade with the CCK1 receptor antagonist tarazepide resulted in approximately 54% suppression of food-stimulated pancreatic enzyme secretion. This finding suggests a significant contribution of CCK1-dependent signalling in piglets, likely mediated predominantly through indirect vagal pathways. However, the relative contribution of direct acinar versus neural mechanisms cannot be resolved from the current data [9,25]. Nevertheless, the demonstration of substantial CCK1-dependent modulation of pancreatic secretion in a large-animal model underscores the potential relevance of CCK-mediated neural pathways in human postprandial pancreatic regulation.

Accordingly, quantitative effects observed in pigs should not be directly extrapolated to humans. Prospective human studies employing analogous experimental designs, together with receptor-selective antagonists and targeted neural interventions, will be required to delineate species-specific regulatory pathways and to refine the translational relevance of these findings.

4.5. Study Limitations

Several limitations of the present study should be acknowledged.

First, although CCK-dependent feedback constituted an important working hypothesis, circulating plasma CCK concentrations did not show statistically significant changes under several experimental conditions. This likely reflects the short plasma half-life of CCK, its pulsatile secretion pattern, and the considerable inter-individual variability characteristic of large-animal models. Moreover, peripheral plasma CCK levels may not accurately reflect local paracrine signalling within the intestinal mucosa, where CCK primarily activates vagal afferent pathways involved in pancreatic exocrine regulation. Accordingly, the absence of significant endocrine CCK responses does not exclude functional involvement of CCK-dependent neural mechanisms.

Second, the observed regulatory effects on pancreatic secretion may involve CCK-independent pathways, including enteric and vagal reflex circuits as well as bile acid–dependent signalling. Although bile acid receptors such as FXR and TGR5 were not directly assessed in the present study, their established roles in enteroendocrine and metabolic regulation suggest that they may contribute to the responses observed following bile diversion and bile acid supplementation. These pathways warrant direct investigation in future mechanistic studies.

Third, trypsin activity in pancreatic juice was assessed using a colorimetric assay, which allows reliable relative comparisons between experimental conditions but offers limited analytical sensitivity at low enzyme activities. Fluorescent substrate–based assays provide improved sensitivity and dynamic range (PMID: 31899136; PMID: 38553043) and may represent a valuable methodological alternative for future studies focusing on protease-dependent feedback mechanisms.

Finally, the relatively small number of animals per experimental condition (n = 6 or fewer) reflects the technical complexity, ethical constraints, and long-term nature of the chronic cannulation model employed. Importantly, each animal served as its own control within a repeated-measures design, substantially reducing inter-individual variability and increasing statistical efficiency despite limited group sizes. Given the small sample size and non-normal distribution of several outcome variables, nonparametric tests for repeated measures (the Friedman test followed by the Wilcoxon signed-rank test) were considered the most appropriate statistical approach and are widely accepted in physiological studies using large-animal models.

No formal correction for multiple comparisons was applied, as the analyses were hypothesis-driven and focused on predefined, physiologically justified contrasts rather than exploratory screening of multiple endpoints. Application of conservative correction procedures under these conditions would markedly increase the risk of type II error and obscure biologically relevant regulatory effects.

The relatively large standard errors observed for some variables reflect inherent biological variability associated with in vivo digestive secretory responses, particularly in large-animal models. Importantly, consistent directional changes across experimental conditions and convergence of multiple independent outcome measures (pancreatic juice volume, protein output, trypsin activity, and bile secretion) support the physiological relevance of the observed effects despite variability in absolute values.

5. Conclusions

This study provides experimental evidence that pancreatic exocrine secretion in weaned piglets is regulated by multiple interacting feedback mechanisms operating along the gastrointestinal tract. Using chronic porcine models, we demonstrate that pancreatic enzyme output is shaped by the integrated action of luminal, hormonal, and bile-related signals.

Protease-dependent luminal feedback emerged as a dominant inhibitory component, as interruption of pancreatic juice flow consistently stimulated enzyme secretion across physiologically relevant protease concentrations and at both proximal and distal intestinal sites. Bile and bile acids acted as modulatory signals, with bile diversion producing sustained stimulation of pancreatic secretion, indicating a tonic inhibitory influence on exocrine output. The specific receptor pathways underlying bile acid–mediated effects remain to be elucidated.

Hormonal pathways involving cholecystokinin and secretin contributed through partially parallel mechanisms, as combined pharmacological blockade resulted in additive suppression of pancreatic secretion. In addition, bile-borne metabolic signals, including leptin, were associated with phase-dependent changes in pancreatic secretion, suggesting a potential modulatory role, although causal relationships were not established.

Overall, the findings support an integrative regulatory framework in which protease-dependent feedback constitutes a core control mechanism, while bile-, hormone-, and metabolism-related signals modulate the magnitude and temporal pattern of pancreatic exocrine secretion.

Author Contributions: Conceptualization, J.L.V.P.; methodology, J.L.V.P.; software, J.L.V.P.; validation, J.L.V.P.; formal analysis, J.L.V.P. and S.E.S.; investigation, J.L.V.P. and S.E.S.; writing—original draft preparation, J.L.V.P.; writing—review and editing, J.L.V.P. and S.E.S.; visualisation, J.L.V.P.; supervision, J.L.V.P.; project administration, J.L.V.P.; funding acquisition, J.L.V.P. All authors have read and agreed to the published version of the manuscript.

Funding: This research was funded by The Swedish Council for Forestry and Agricultural Research, Visby programme, grants no. 7391/1998/380/67 and 2113/1999/380/71; partial funding Solvay Pharmaceuticals GmbH (Germany).

Institutional Review Board Statement: The animal study protocol was approved by the local Ethics Committee of Lund University (Sweden) (permission no. M238-99). The research was conducted on weaned piglets of a commercial crossbreed (Large White Swedish/Hampshire × Yorkshire) obtained

https://doi.org/10.3390/biom16020322

from the experimental farm in Odarslöv belonging to the Swedish University of Agricultural Sciences (Alnarp, Sweden).

Informed Consent Statement: Not applicable.

Data Availability Statement: The data supporting the findings of this study are available from the corresponding author upon reasonable request. Public access to the data is restricted due to ethical approval conditions and institutional policies regarding animal experimentation.

Acknowledgments: During the preparation of this manuscript, the authors used ChatGPT (OpenAI, GPT-5.2) to support language editing and stylistic refinement of the text, as well as the development of the pig model presented in Figure 2. In addition, BioRender was used to create the graphical abstract [55] and all Figures [21,30]. The authors reviewed and edited all generated content and take full responsibility for the integrity and accuracy of this publication. Thanks to Solvay Pharmaceuticals GmbH (Germany) for providing the CCK1 receptor antagonist (tarazepide) and a preparation containing active pancreatic enzymes (pancreatin). Lund University, Sweden, for providing animal experiments and performing enzymatic analyses and RIAs. University of Life Sciences in Lublin for supporting the realisation of the studies. The authors thank Dominika Nguyen Ngoc for her assistance in the preparation of the figures.

Conflicts of Interest: The authors declare no conflicts of interest.

Abbreviations

The following abbreviations are used in this manuscript:

ATP	Adenosine Triphosphate
BAPNA	Nα-benzoyl-DL-arginine p-nitroanilide
BSA	Bovine Serum Albumin
cAMP	Cyclic Adenosine Monophosphate
CCK	Cholecystokinin
CCK-33	CCK Octapeptide
CCK1	Peripheral Cholecystokinin Receptor (CCK-A; A—Alimentary)
CCK2	Central Cholecystokinin Receptor (CCK-B; B—Brain)
CGRP	Calcitonin Gene-Related Peptide
EDTA	Ethylenediaminetetraacetic Acid
GIP	Gastric Inhibitory Peptide
GLP 1–36	Glucagon-Like Peptide 1–36 (Glucagon-like Growth Factor I)
GLP2	Glucagon-Like Peptide 2 (Glucagon-like Growth Factor II)
GRP	Gastrin-Releasing Peptide
H+/K+-ATPase	Proton-potassium pump adenosine triphosphatase
id	Intraduodenal
IgG	Immunoglobulin G
il	Intraileal
K_3EDTA	Tripotassium ethylenediaminetetraacetate
iv	Intravenous
I-cells	CCK-producing cells
b.w.	Body weight
MMC	Migrating Motor Complex
MP	Monitor Peptide
mRNA	messenger Ribonucleic Acid
n	Number of animals in the experiment
PP	Pancreatic Polypeptide
PYY	Pepetide YY
PYY 1–36	Peptide YY fragment 1–36
PYY 3–36	Peptide YY fragment 3–36
SE	Standard Error

SRP	Secretin-Releasing Peptides
VIP	Vasoactive Intestinal Peptide
PJ	Pancreatic Juice
PEG	Pegylated Leptin
LCRF	Luminal CCK-Releasing Factor
PLC	Phospholipase C
NTS	Nucleus Tractus Solitarius
AC6	Adenylyl Cyclase 6
SCTR	Secretin Receptors
PKA	Protein Kinase A
PKC	Protein Kinase C
IP_3R	Inositol 1,4,5-triphosphate Receptor
FXR	Farnesoid X Receptor
NR1H4	Nuclear Receptor Subfamily 1 Group H Member 4
CDCA	Chenodeoxycholic Acid
FGF15/19	Fibroblast Growth Factor 15/19
CYP7A1	Cholesterol 7α-hydroxylase
TGR5	Takeda G Protein-Coupled Receptor 5 (GPBAR1)
POMC	Pro-opiomelanocortin
B	Bile
U	Unit

Appendix A

Appendix A.1

Table A1. Scheme of experiments 1–14 performed using experimental model I.

Experimental Options	Experiment Number													
	1	2	3	4	5	6	7	8	9	10	11	12	13	14
Pancreatic juice infusions into duodenum	+	+	+	+	-	-	-	-	+	+	+	-	-	-
Pancreatin infusions into duodenum	-	20%	100%	200%	-	20%	100%	200%	-	-	-	-	-	-
Pancreatin infusions into ileum	-	-	-	-	-	-	-	-	20%	100%	200%	20%	100%	200%

Legend: + administration of pancreatic juice to the duodenum in amounts equal to those secreted under control into the duodenum. - no administration of pancreatic juice or pancreatin to the duodenum or to the ileum. 20% infusion of pancreatin solution with trypsin activity corresponding to 20% of trypsin activity during the feeding period. 100% infusion of pancreatin solution with trypsin activity corresponding to trypsin activity during the feeding period. 200% infusion of pancreatin solution with trypsin activity corresponding to double the trypsin activity during the feeding period.

Table A2. Scheme of experiments 15–21 performed using model II with administration of omeprazole, tarazepide, and bicarbonates.

Experiment Number	Experiment Options				
	Pancreatic Juice	Bile	Omeprazole	Tarazepide	Bicarbonates *
15	+	+	-	-	-
16	+	-	-	-	-
17	-	-	-	-	-
18	+	+	+	-	-
19	+	+	-	+	-
20	+	+	+	+	-
21	+	-	-	-	+

Legend: + Administration of pancreatic juice, bile, bicarbonate, omeprazole, or tarazepide during the experiment. - No administration of pancreatic juice, bile, bicarbonate, omeprazole, or tarazepide during the experiment. * Bicarbonate was administered to the duodenum in amounts corresponding to the volume of bile secreted in the 30 min period of bile collection immediately prior to its administration.

https://doi.org/10.3390/biom16020322

Table A3. Scheme of experiments 22–26 performed using model II with preserved return of bile and pancreatic juice to the duodenum and with interruption of their flow to the duodenum, as well as with infusions of bile acid salt solutions to the duodenum.

Experiment Number	Experiment Options			
	Pancreatic Juice	Bile During 10 h Before Experiments	Bile During Experiments	Bile Acids Salts Solutions
22	+	+	+	-
23	+	-	-	-
24	+	-	-	0.02 M
25	+	-	-	0.04 M
26	+	-	-	0.08 M

Legend: + Administration of pancreatic juice or bile to the duodenum before or during the experiment. - No administration of bile to the duodenum before or during the experiment. 0.02 M; 0.04 M; 0.08 M Molarity of bile acid salt solutions administered to the duodenum during the feeding phase and in the first hour of the post-feeding phase.

Appendix A.2. Protocol Design of Experiments Showing Sample Collection Time and Infusion Time

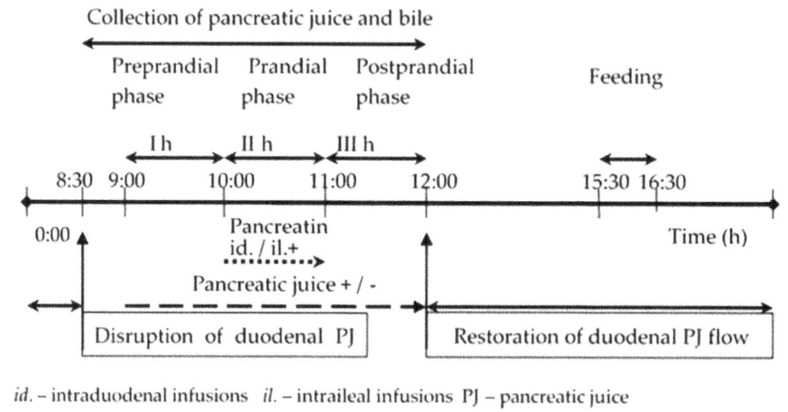

Scheme A1. Exp. 1–14.

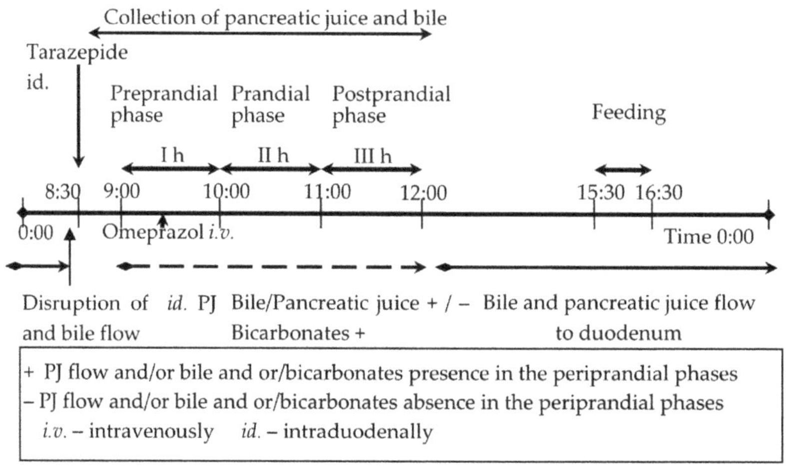

Scheme A2. Exp. 15–21.

https://doi.org/10.3390/biom16020322

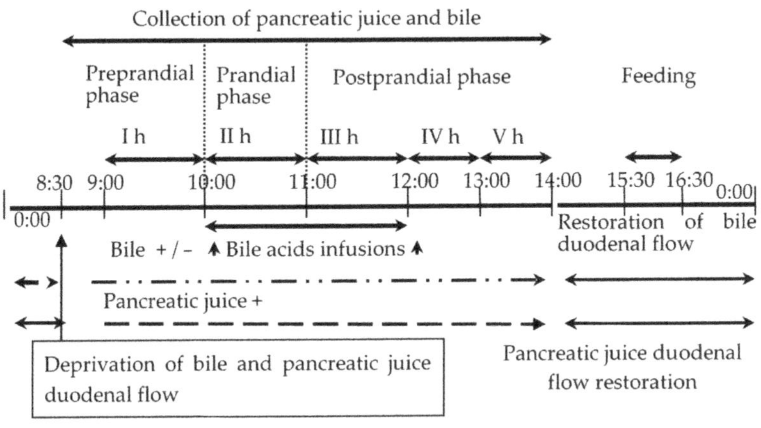

Scheme A3. Exp. 22–26.

References

1. Chandra, R.; Liddle, R.A. Neural and Hormonal Regulation of Pancreatic Secretion. *Curr. Opin. Gastroenterol.* **2009**, *25*, 441–446. [CrossRef]
2. Chandra, R.; Liddle, R.A. Recent Advances in the Regulation of Pancreatic Secretion. *Curr. Opin. Gastroenterol.* **2014**, *30*, 490–494. [CrossRef] [PubMed]
3. Roura, E.; Koopmans, S.-J.; Lallès, J.-P.; Le Huerou-Luron, I.; de Jager, N.; Schuurman, T.; Val-Laillet, D. Critical Review Evaluating the Pig as a Model for Human Nutritional Physiology. *Nutr. Res. Rev.* **2016**, *29*, 60–90. [CrossRef] [PubMed]
4. Owyang, C.; Louie, D.S.; Tatum, D. Feedback Regulation of Pancreatic Enzyme Secretion. Suppression of Cholecystokinin Release by Trypsin. *J. Clin. Investig.* **1986**, *77*, 2042–2047. [CrossRef] [PubMed]
5. Louie, D.S.; May, D.; Miller, P.; Owyang, C. Cholecystokinin Mediates Feedback Regulation of Pancreatic Enzyme Secretion in Rats. *Am. J. Physiol.* **1986**, *250*, G252–G259. [CrossRef]
6. Miyasaka, K.; Guan, D.F.; Liddle, R.A.; Green, G.M. Feedback Regulation by Trypsin: Evidence for Intraluminal CCK-Releasing Peptide. *Am. J. Physiol.* **1989**, *257*, G175–G181. [CrossRef]
7. Lu, L.; Louie, D.; Owyang, C. A Cholecystokinin Releasing Peptide Mediates Feedback Regulation of Pancreatic Secretion. *Am. J. Physiol.* **1989**, *256*, G430–G435. [CrossRef]
8. Murphy, J.A.; Criddle, D.N.; Sherwood, M.; Chvanov, M.; Mukherjee, R.; McLaughlin, E.; Booth, D.; Gerasimenko, J.V.; Raraty, M.G.T.; Ghaneh, P.; et al. Direct Activation of Cytosolic Ca^{2+} Signaling and Enzyme Secretion by Cholecystokinin in Human Pancreatic Acinar Cells. *Gastroenterology* **2008**, *135*, 632–641. [CrossRef]
9. Morisset, J.; Levenez, F.; Corring, T.; Benrezzak, O.; Pelletier, G.; Calvo, E. Pig Pancreatic Acinar Cells Possess Predominantly the CCK-B Receptor Subtype. *Am. J. Physiol.* **1996**, *271*, E397–E402. [CrossRef]
10. Sabbatini, M.E.; D'Alecy, L.; Lentz, S.I.; Tang, T.; Williams, J.A. Adenylyl Cyclase 6 Mediates the Action of Cyclic AMP-Dependent Secretagogues in Mouse Pancreatic Exocrine Cells via Protein Kinase A Pathway Activation. *J. Physiol.* **2013**, *591*, 3693–3707. [CrossRef]
11. Humbert, L.; Rainteau, D.; Tuvignon, N.; Wolf, C.; Seksik, P.; Laugier, R.; Carrière, F. Postprandial Bile Acid Levels in Intestine and Plasma Reveal Altered Biliary Circulation in Chronic Pancreatitis Patients. *J. Lipid Res.* **2018**, *59*, 2202–2213. [CrossRef]
12. Santos, L.G.A.-A.; Musther, H.; Bala, N.; Deferm, N.; Patel, G.; Brouwers, J.; Turner, D.B. Gastrointestinal Bile Salt Concentrations in Healthy Adults Under Fasted and Fed Conditions: A Systematic Review and Meta-Analysis for Mechanistic Physiologically-Based Pharmacokinetic (PBPK) Modelling. *AAPS J.* **2025**, *27*, 31. [CrossRef] [PubMed]
13. Inagaki, T.; Choi, M.; Moschetta, A.; Peng, L.; Cummins, C.L.; McDonald, J.G.; Luo, G.; Jones, S.A.; Goodwin, B.; Richardson, J.A.; et al. Fibroblast Growth Factor 15 Functions as an Enterohepatic Signal to Regulate Bile Acid Homeostasis. *Cell Metab.* **2005**, *2*, 217–225. [CrossRef] [PubMed]
14. Thomas, C.; Gioiello, A.; Noriega, L.; Strehle, A.; Oury, J.; Rizzo, G.; Macchiarulo, A.; Yamamoto, H.; Mataki, C.; Pruzanski, M.; et al. TGR5-Mediated Bile Acid Sensing Controls Glucose Homeostasis. *Cell Metab.* **2009**, *10*, 167–177. [CrossRef] [PubMed]
15. Pierzynowski, S.; Sileikiene, V.; Valverde Piedra, J.L.; Szymanczykc, S.; Gregory, P.C.; Kruszewska, D.; Mosenthin, R.; Rzasa, A.; Kowalik, S.; Zabielski, R.; et al. Real Exposure to Pig Pancreatic Juice and Bile Inhibit Exocrine Pancreatic Secretion in Pigs. *Livest. Sci.* **2007**, *108*, 53–56. [CrossRef]

https://doi.org/10.3390/biom16020322

16. Valverde Piedra, J.L.; Pierzynowski, S. Attempt to Establish a Chronic Model to Study the Influence of Bile and Pancreatic Juice Diversion on Pancreas Feedback Regulation in Conscious Pigs. *Livest. Sci.* **2007**, *109*, 46–49. [CrossRef]
17. Riepl, R.L.; Lehnert, P. The Mediators of Bile Action on the Exocrine Pancreas. *Scand. J. Gastroenterol.* **1993**, *28*, 369–374. [CrossRef]
18. Guilmeau, S.; Nagain-Domaine, C.; Buyse, M.; Tsocas, A.; Rozé, C.; Bado, A. Modulation of Exocrine Pancreatic Secretion by Leptin through CCK(1)-Receptors and Afferent Vagal Fibres in the Rat. *Eur. J. Pharmacol.* **2002**, *447*, 99–107. [CrossRef]
19. Matyjek, R.; Herzig, K.-H.; Kato, S.; Zabielski, R. Exogenous Leptin Inhibits the Secretion of Pancreatic Juice via a Duodenal CCK1-Vagal-Dependent Mechanism in Anaesthetized Rats. *Regul. Pept.* **2003**, *114*, 15–20. [CrossRef]
20. Midha, S.; Singh, S.; Sachdev, V.; Misra, A.; Garg, P.K. Leptin and Its Correlation with Exocrine and Endocrine Pancreatic Function in Idiopathic Chronic Pancreatitis: Implications for Pathophysiology. *Pancreas* **2007**, *35*, 262–266. [CrossRef]
21. Nguyen Ngoc, D. Figure 1. Available online: https://BioRender.com/npb2j28 (accessed on 17 February 2026).
22. Pierzynowski, S.G.; Weström, B.R.; Svendsen, J.; Karlsson, B.W. Development of Exocrine Pancreas Function in Chronically Cannulated Pigs during 1–13 Weeks of Postnatal Life. *J. Pediatr. Gastroenterol. Nutr.* **1990**, *10*, 206–212. [CrossRef]
23. Thaela, M.J.; Pierzynowski, S.G.; Jensen, M.S.; Jakobsen, K.; Weström, B.R.; Karlsson, B.W. The Pattern of the Circadian Rhythm of Pancreatic Secretion in Fed Pigs. *J. Anim. Sci.* **1995**, *73*, 3402–3408. [CrossRef] [PubMed]
24. Rantzer, D.; Kiela, P.; Thaela, M.J.; Svendsen, J.; Ahrén, B.; Karlsson, S.; Pierzynowski, S.G. Pancreatic Exocrine Secretion during the First Days after Weaning in Pigs. *J. Anim. Sci.* **1997**, *75*, 1324–1331. [CrossRef] [PubMed]
25. Zabielski, R.; Leśniewska, V.; Borlak, J.; Gregory, P.C.; Kiela, P.; Pierzynowski, S.G.; Barej, W. Effects of Intraduodenal Administration of Tarazepide on Pancreatic Secretion and Duodenal EMG in Neonatal Calves. *Regul. Pept.* **1998**, *78*, 113–123. [CrossRef] [PubMed]
26. Cederberg, C.; Röhss, K.; Lundborg, P.; Olbe, L. Effect of Once Daily Intravenous and Oral Omeprazole on 24-Hour Intragastric Acidity in Healthy Subjects. *Scand. J. Gastroenterol.* **1993**, *28*, 179–184. [CrossRef]
27. Haven, M.L.; Dave, K.; Burrow, J.A.; Merritt, A.M.; Harris, D.; Zhang, D.; Hickey, G.J. Comparison of the Antisecretory Effects of Omeprazole When Administered Intravenously, as Acid-Stable Granules and as an Oral Paste in Horses. *Equine Vet. J. Suppl.* **1999**, *29*, 54–58. [CrossRef]
28. Erlanger, B.F.; Kokowsky, N.; Cohen, W. The Preparation and Properties of Two New Chromogenic Substrates of Trypsin. *Arch. Biochem. Biophys.* **1961**, *95*, 271–278. [CrossRef]
29. Döring, S.; Wulfes, B.S.; Atanasova, A.; Jaeger, C.; Walzel, L.; Tscheuschner, G.; Flemig, S.; Gawlitza, K.; Feldmann, I.; Konthur, Z.; et al. Corundum Particles as Trypsin Carrier for Efficient Protein Digestion. *BioTech* **2025**, *15*, 2. [CrossRef]
30. Nguyen Ngoc, D. Figure 2. Available online: https://BioRender.com/f4pcgw1 (accessed on 17 February 2026).
31. Chandra, R.; Liddle, R.A. Modulation of Pancreatic Exocrine and Endocrine Secretion. *Curr. Opin. Gastroenterol.* **2013**, *29*, 517–522. [CrossRef]
32. Chey, W.Y.; Chang, T. Neural Hormonal Regulation of Exocrine Pancreatic Secretion. *Pancreatology* **2001**, *1*, 320–335. [CrossRef]
33. Owyang, C.; Logsdon, C.D. New Insights into Neurohormonal Regulation of Pancreatic Secretion. *Gastroenterology* **2004**, *127*, 957–969. [CrossRef] [PubMed]
34. Kuhre, R.E.; Wewer Albrechtsen, N.J.; Larsen, O.; Jepsen, S.L.; Balk-Møller, E.; Andersen, D.B.; Deacon, C.F.; Schoonjans, K.; Reimann, F.; Gribble, F.M.; et al. Bile Acids Are Important Direct and Indirect Regulators of the Secretion of Appetite- and Metabolism-Regulating Hormones from the Gut and Pancreas. *Mol. Metab.* **2018**, *11*, 84–95. [CrossRef] [PubMed]
35. Chen, X.-L.; Xie, K.-X.; Yang, Z.-L.; Yuan, L.-W. Expression of FXR and HRG and Their Clinicopathological Significance in Benign and Malignant Pancreatic Lesions. *Int. J. Clin. Exp. Pathol.* **2019**, *12*, 2111–2120. [PubMed]
36. Chiang, J.Y.L. Bile Acid Metabolism and Signaling. *Compr. Physiol.* **2013**, *3*, 1191–1212. [CrossRef]
37. Rehfeld, J.F. The Molecular Nature of Cholecystokinin in Plasma: An in Vivo Immunosorption Study in Rabbits. *Scand. J. Gastroenterol.* **1994**, *29*, 110–121. [CrossRef]
38. Zabielski, R.; Piedra, J.L.V.; Gacsalyi, U.; Laubitz, D.; Jakob, S.; Rzasa, A.; Gregory, P.C.; Pierzynowski, S.G. Influence of Luminal Pancreatic Juice and Bile on the Secretion of Pancreatic Juice and Bile in Weaned Pigs. In *Digestive Physiology of Pigs. Proceedings of the 8th Symposium, Swedish University of Agricultural Sciences, Uppsala, Sweden, 20–22 June 2000*; CABI Books: Wallingford, UK, 2001; pp. 218–220. ISBN 978-0-85199-517-5.
39. Kuzmina, I.V.; Tolpygo, S.M.; Kotov, A.V.; Shoibonov, B.B.; Zamolodchikova, T.S. Basal Pancreatic Secretion in a Comparative Aspect in Poultry and Rodents. *Front. Physiol.* **2024**, *15*, 1340130. [CrossRef]
40. Kong, W.; McConalogue, K.; Khitin, L.M.; Hollenberg, M.D.; Payan, D.G.; Böhm, S.K.; Bunnett, N.W. Luminal Trypsin May Regulate Enterocytes through Proteinase-Activated Receptor 2. *Proc. Natl. Acad. Sci. USA* **1997**, *94*, 8884–8889. [CrossRef]
41. Liddle, R.A. Regulation of Cholecystokinin Secretion by Intraluminal Releasing Factors. *Am. J. Physiol.-Gastrointest. Liver Physiol.* **1995**, *269*, G319–G327. [CrossRef]
42. Koop, I. Role of Bile Acids in the Control of Pancreatic Secretion and CCK Release. *Eur. J. Clin. Investig.* **1990**, *20*, S51–S57. [CrossRef]

https://doi.org/10.3390/biom16020322

43. Nawrot-Porabka, K.; Jaworek, J.; Leja-Szpak, A.; Palonek, M.; Szklarczyk, J.; Konturek, S.J.; Pawlik, W.W. Leptin Is Able to Stimulate Pancreatic Enzyme Secretion via Activation of Duodeno-Pancreatic Reflex and CCK Release. *J. Physiol. Pharmacol.* **2004**, *55*, 47–57.
44. Takahashi, T. Interdigestive Migrating Motor Complex—Its Mechanism and Clinical Importance. *J. Smooth Muscle Res.* **2013**, *49*, 99–111. [CrossRef] [PubMed]
45. Claudel, T.; Sturm, E.; Duez, H.; Torra, I.P.; Sirvent, A.; Kosykh, V.; Fruchart, J.-C.; Dallongeville, J.; Hum, D.W.; Kuipers, F.; et al. Bile Acid-Activated Nuclear Receptor FXR Suppresses Apolipoprotein A-I Transcription via a Negative FXR Response Element. *J. Clin. Investig.* **2002**, *109*, 961–971. [CrossRef] [PubMed]
46. Garces, M.C.; Gomez-Cerezo, J.; Codoceo, R.; Grande, C.; Barbado, J.; Vazquez, J.-J. Postprandial Cholecystokinin Response in Patients with Chronic Pancreatitis in Treatment with Oral Substitutive Pancreatic Enzymes. *Dig. Dis. Sci.* **1998**, *43*, 562–566. [CrossRef] [PubMed]
47. Eddes, E.H.; Masclee, A.A.; Gielkens, H.A.; Verkijk, M.; Vecht, J.; Biemond, I.; Lamers, C.B. Cholecystokinin Secretion in Patients with Chronic Pancreatitis and after Different Types of Pancreatic Surgery. *Pancreas* **1999**, *19*, 119–125. [CrossRef]
48. Chiang, J.Y.L.; Pathak, P.; Liu, H.; Donepudi, A.; Ferrell, J.; Boehme, S. Intestinal FXR and TGR5 Signaling in Metabolic Regulation. *Dig. Dis.* **2017**, *35*, 241–245. [CrossRef]
49. Berglund, E.D.; Vianna, C.R.; Donato, J.; Kim, M.H.; Chuang, J.-C.; Lee, C.E.; Lauzon, D.A.; Lin, P.; Brule, L.J.; Scott, M.M.; et al. Direct Leptin Action on POMC Neurons Regulates Glucose Homeostasis and Hepatic Insulin Sensitivity in Mice. *J. Clin. Investig.* **2012**, *122*, 1000–1009. [CrossRef]
50. Ziomek, M.; Drozd, Ł.; Chałabis-Mazurek, A.; Szkucik, K.; Paszkiewicz, W.; Valverde Piedra, J.L.; Bełkot, Z.; Maćkowiak-Dryka, M.; Gondek, M.; Knysz, P. Concentration Levels of Cadmium and Lead in the Raw and Processed Meat of Helix Pomatia Snails. *Pol. J. Vet. Sci.* **2018**, *21*, 483–489. [CrossRef]
51. Medhus, A.W.; Sandstad, O.; Näslund, E.; Hellström, P.M.; Husebye, E. The Influence of the Migrating Motor Complex on the Postprandial Endocrine Response. *Scand. J. Gastroenterol.* **1999**, *34*, 1012–1018. [CrossRef]
52. Owyang, C.; Achem-Karam, S.R.; Vinik, A.I. Pancreatic Polypeptide and Intestinal Migrating Motor Complex in Humans: Effect of Pancreatic Obiliary Secretion. *Gastroenterology* **1983**, *84*, 10–17. [CrossRef]
53. Evilevitch, L.; Weström, B.R.; Pierzynowski, S.G. CCK Regulates Pancreatic Enzyme Secretion via Short Duodenal-Pancreatic Reflexes in Pigs. *Scand. J. Gastroenterol.* **2003**, *38*, 201–206. [CrossRef]
54. Dufresne, M.; Seva, C.; Fourmy, D. Cholecystokinin and Gastrin Receptors. *Physiol. Rev.* **2006**, *86*, 805–847. [CrossRef]
55. Nguyen Ngoc, D. Graphical Abstract. Available online: https://BioRender.com/rsegok3 (accessed on 17 February 2026).

MDPI AG
Grosspeteranlage 5
4052 Basel
Switzerland
Tel.: +41 61 683 77 34

Biomolecules Editorial Office
E-mail: biomolecules@mdpi.com
www.mdpi.com/journal/biomolecules

www.ingramcontent.com/pod-product-compliance
Lightning Source LLC
Chambersburg PA
CBHW062101220926
51850CB00010B/2848